W9-CBC-333

Additional Praise for *Green*:

"Michael and Jane have come up with a uniquely personal account of today's energy revolution. It unravels the complexities of renewable energy and gives shape to today's policy debate."

—Lord John Browne, retired CEO of BP

"Jane and Michael Hoffman have hit a bulls-eye with their rich mandala of a book, *Green*. Profoundly ambitious and packed with facts, figures and fancy, the book is at once an instructive history lesson and an effective 'greenprint' for the future. For our global environmental ills, the Hoffmans wisely prescribe an antidote of informed individual action complemented and facilitated by enlightened public policy."

—Rob Watson, Chairman, EcoTech International, Founding Chair, LEED Steering Committee

"*Green* is an exceptionally relevant, timely, and informative book. In clear and concise prose, Jane and Michael Hoffman offer critical understanding for the average person of the complexities of energy use and policy from the local to the global level. This book brims with inspirational guidance and, above all, optimistic and hopeful knowledge that all of us can meaningfully incorporate into our lives."

—Stephen R. Kellert, Tweedy Ordway Professor of Social Ecology, Yale University, School of Forestry and Environmental Studies; Partner, Environmental Capital Partners

GREEN

GREEN

YOUR PLACE IN THE NEW ENERGY REVOLUTION

JANE HOFFMAN

AND

MICHAEL HOFFMAN

palgrave
macmillan

GREEN
Copyright © Jane Hoffman and Michael Hoffman, 2008.

First published in 2008 by
PALGRAVE MACMILLAN™
175 Fifth Avenue, New York, N.Y. 10010 and
Houndmills, Basingstoke, Hampshire, England RG21 6XS
Companies and representatives throughout the world.

PALGRAVE MACMILLAN is the global academic imprint of the Palgrave
Macmillan division of St. Martin's Press, LLC and of Palgrave Macmillan Ltd.
Macmillan® is a registered trademark in the United States, United Kingdom
and other countries. Palgrave is a registered trademark in the European
Union and other countries.

ISBN-13: 978–0–230–60544–2
ISBN-10: 0–230–60544–3

Library of Congress Cataloging-in-Publication Data

Hoffman, Jane S., 1964–
    Green : your place in the new energy revolution / Jane Hoffman and
Michael Hoffman.
        p. cm.
    Includes bibliographical references and index.
    ISBN 0–230–60544–3
        1. Environmental policy—United States—Citizen participation.
    2. Environmental protection—United States—Citizen participation.
    3. Renewable energy sources—United States. 4. Green movement—
United States. I. Hoffman, Michael B., 1950– II. Title.

HC110.E5H625 2008
333.799′4—dc22   2007052853

A catalogue record for this book is available from the British Library.

Design by Newgen Imaging Systems (P) Ltd., Chennai, India.

First edition: July 2008

10 9 8 7 6 5 4 3 2 1

Printed in the United States of America.

CONTENTS

"Not only will atomic power be released, but someday we will harness the rise and fall of tides and imprison the rays of the sun."

—Thomas Edison

GREEN

Introduction

**EVEN DOGS CAN
SEE IN GREEN**

LIKE ANY OTHER RESPONSIBLE DOG OWNER, WHEN WE TAKE OUR WHEATON terrier, Liberty, out for her daily walk, we make sure that one of us has a plastic bag tucked into a pocket to use for cleanup duty. In our environmentally conscious world, however, the simple act of curbing your dog can cause an alert person to pause: Plastic, because it is impermeable, is the most convenient and logical material to use for the messy job of picking up after your dog. Plastic is also a petroleum product. Producing it releases carbon dioxide into the environment and contributes to the accumulation of greenhouse gases in the atmosphere. Moreover, our best science estimates that it will take anywhere from 500 to 1,000 years for discarded plastic to break down to its elements, and even then the individual, man-made molecules of plastic that remain won't be recognized as food by the microorganisms necessary to complete the cycle of returning the elements to the earth. By maintaining a clean public sidewalk, you are essentially putting your dog's poop into a container that will preserve it in some landfill into the next millennium.

It's unlikely that there are many people who haven't yet heard the inconvenient truth that our planet is facing its eleventh hour, who haven't been engaged in the global conversation about global warming or been caught up to some degree in the anxiety to *do something about it*.

But what? If even the everyday task of picking up after your dog contributes to the problem, how can any individual hope to be a part of the larger solution? Can changing the sort of light bulbs you use, putting a blanket on your water heater, and unplugging your computer when it's not in use possibly be enough? Are more drastic measures called for? Do we all have to start raising our own livestock, install composting toilets, become vegan in order to assure that our children inherit a healthy planet? If so, what does that mean for the quality of life we have all come to enjoy? Suppose you do trade in your SUV for a hybrid car; is it realistic to expect that this small contribution will offset the pollution that's being generated down the road, at your friendly neighborhood coal-fired power plant? That it will in any way mitigate the carbon emissions being manufactured in Asia, as China builds a new city the size of Philadelphia each and every month? It's not hard to feel overwhelmed by the urgency of the environmental message or defeated in trying to sort through the sometimes misleading information that's presented in the media as fact. Indeed, *panic* isn't too strong a word to describe the state of mind of some of the people we've spoken with about the earth's future.

Panic, however, won't lead us to thoughtful solutions that are useful to us now and that will sustain a livable planet for the generations who will follow us.

That's what this book is about: solutions.

The first step in determining the most useful ones is to take a step back. Take a breath. Put the problem into perspective.

The problem, of course, is real enough. It's unreasonable, if not downright irresponsible, to argue with the overwhelming consensus within the scientific community that tells us that levels of carbon dioxide (CO_2) in the earth's atmosphere are rising and compromising our planet's climate. We are, after all, experiencing the results of the compromise firsthand in record-setting temperatures, the increasing severity of droughts, and the increasing violence of other natural phenomena, like hurricanes and tornadoes. We deal almost daily with reminders of the changes in our global weather patterns whether we live in a tsunami-devastated part of Thailand or a not-yet-rebuilt part of New Orleans—or if we merely see heartbreaking footage of these places on the nightly news, prompting us to write another check to another relief agency.

Individuals—you and I—are acutely aware there is a problem. We'd be hard-pressed to pick up a newspaper and not find at least one or two articles about energy conservation, alternative fuels, or renewable power sources. Ask any fourth grader and she'll probably be able to boil down for you in schoolchild language the sophisticated opinions of the world's top scientists: The rise of CO_2 in the earth's atmosphere is directly linked to our dependence on burning the fossil fuels oil, coal, and natural gas to create the energy we need. But this sense of urgency is also the one truly positive outcome of fostering that underlying, rumbling sense of panic about the environment: People, you and I, *get it*. One by one we've been breaking out of our comfort zones, facing the limitations of our traditional sources of power and fuel, recognizing that our growing need for energy is enormous. You and I understand that it's impractical, if not dangerous, to stick our heads in the sand about finding solutions for our energy future, and we've been stepping up the pressure on leaders who have been frustratingly slow to address the problem with realism and ingenuity. Our energy challenges become less and less abstract as more people are reaching their *Aha!* moments about the issue more quickly and positioning themselves to act on it.

Michael and I had a simultaneous *Aha!* moment. Probably like yours, ours didn't strike out of the clear blue sky; it was a culmination of events and information gathering. Like yours too, ours wasn't the result of a sign of biblical proportions, but something much more humble—and, for us, just as unmistakable: a tuna sandwich.

In 2000, while I was Commissioner of the New York City Department of Consumer Affairs, I was working on a lawsuit that involved a line of sightseeing tourist buses. The buses' emissions weren't meeting current EPA standards. While I was researching the composition of the particulates that these buses were putting out into the air for New Yorkers to breathe—the legal and scientific aspects of the lawsuit—what kept nagging at me was the moral aspect of the case, the moral imperative of the bus company to conduct its business responsibly. Why wasn't the company using the newer-model buses that were fueled by natural gas, or maybe diesel hybrid-electric buses, for its fleet? Those sorts of buses would allow the company to operate a viable tourist business at the same time as it was acting as a good New York neighbor.

The foundation for Michael's energy *Aha!* was his long involvement in the energy business. Michael began his career in 1973 at the time of the first oil price shock, working within U.S. federal agencies to reallocate energy resources and resolve huge inefficiencies in the regulatory system. In the private sector, his focus has long been on energy and energy investment. In 2001 he was a consultant to Governor Gray Davis during California's energy emergency, and later, during the Enron bankruptcy, he'd been in charge of selling off Enron's assets. Now he runs a private equity fund in the renewable energy area. Because Michael is a businessman himself, he knows that in order for business to change the way it does business there has to be an economic incentive. This is especially true in the case of meeting our energy challenge because the scale of change we're talking about is big. Michael understood what was happening at the epicenter of the traditional energy market—the gathering forces of increasing demand, decreasing supplies of fossil fuels, expanding awareness of the toll our energy appetite was taking on the environment—so he knew early on that a big change was going to have to come.

Scientists and engineers had already, of course, been developing renewable energy technologies. Some of their ideas were brilliant; others were, to put it kindly, less than viable. What Michael did was to start sorting prudently through the technologies for this emerging marketplace to determine which held the most potential. If there were going to be buyers for a new kind of power—and, indeed, there were going to *have* to be—renewable energy was going to have to be profitable.

It was in the course of investing in one of his first renewable energy projects, a utility scale solar field, that Michael found out just how enthusiastically key investors were responding to the idea of renewable energy as a new area for private equity. They *wanted* to invest in this stuff. Renewable energy was a classic platform to do well by doing good.

And then, one afternoon, I served tuna fish for lunch.

Tuna fish is one of Michael's favorites so it's a happy coincidence that, given my limited repertoire, it is also one of my specialties. But tuna had also become a point of contention between us. Michael's company owned a coal-fired power plant, and such coal-fired plants emit, among other pollutants, mercury. Mercury emissions settle in bodies of water, or on land where they are eventually flushed into bodies of water: the tuna's habitat. In response to a spate of reports about health risks associated with eating mercury-poisoned fish, I'd been rationing my family's consumption.

That afternoon, while Michael and I were sitting in the nook off our kitchen, enjoying our lunch together, he innocently complimented me on the sandwiches. I took it as another opportunity to press my point about the coal-fired plant.

"The tuna looked too fresh at the market to pass it up," I replied. "Too bad we can't have it more than once a week. I guess we have you to thank for that."

I don't recall that Michael and I exchanged another word about tuna. We didn't need to. We had made it to our *Aha!*, that tipping point between intellectually knowing what the problem is and emotionally connecting to the need to be a part of solving it. Michael ultimately sold the coal-fired plant and launched what is now the world's largest investment fund for renewable energy.

Let's take stock: We have an increasingly informed public motivated to implement solutions to our energy challenge. We have the moral imperative to steward our earth so that we—and the generations who come after us—can enjoy life in a healthy environment. And we have an economic incentive for investment in technology that will sustainably meet our energy needs.

The piece of the puzzle that remains most important is the informed public—getting the right information about reasonable solutions into the hands of the people. This piece is critical because, like most meaningful revolutions, the green revolution is a "bottom-up" movement. Like the American Revolution that created a brand-new country out of a wilderness, or the telecommunications revolution that has in the last ten or twenty years redefined the way the world does business, the energy revolution is the result of informed, caring people forcing their leaders to lead.

In order to know what to tell our leaders to do, we've got to put aside the nagging sense of urgency and cut thoughtfully through the confusion of facts and misinformation. This book is intended to be a tool to help in that process.

Our focus in this book is on renewable energy and the part it will play in meeting our energy challenge. We'll define what renewable energy is, the pros and cons of each type of renewable energy, and how they are being successfully incorporated as part of state, national, and international energy programs.

But while renewable energy is a big part of the solution, it is only one part of a three-part formula we need to implement to create sustainable energy policies.

Renewable Energy + Increased Energy Conservation + Energy Technologies that Don't Exist Yet = A Secure Energy Future

Increased energy conservation means that we need to become aware of the ways in which we use the energy that we already have. Changing the sort of light bulbs you use in the lamps in your living room, putting a blanket over your water heater, and turning off your computer when you're not using it are all meaningful steps you can take to have an immediate, positive impact on our energy reserves.

In chapter 8, we'll talk about what we mean by energy technologies that don't exist yet—some of the innovations that are now firing the imaginations of countries and corporations, scientists and engineers, inventors and *investors*.

But no man is an island. As humans require the intricate web of relationships that form our complex society and allow it to function efficiently—and often pleasurably!—none of the parts of this formula can stand alone. It's exactly their interconnectedness, in fact, that inspires not merely optimism that the problem of our global energy needs is manageable, but real excitement at the abundance of opportunities we have to transform our world for the better.

Our monetary system was originally founded on the gold standard—that is, our paper money got its worth from the gold reserves that gave it real-world value. Taken together, the three parts of our formula can create a Green Standard—a platform to determine and implement the best practices in the ways we use and invest in energy so that we enhance and sustain the real-world value of our planet's resources.

We believe the Green Standard is the future. We believe this because once any responsible individual, once any patriotic citizen, once any concerned member of the global community takes a step toward going green, not one step is taken backward anywhere in the world.

In this book we'll talk about how energy independence is vital to the security of nations now dependent on foreign oil—and we'll talk about it not in the language of sacrifice, but in terms of how current research and development can mean a revival of domestic economies based on a whole new industry and the green-collar jobs it will create.

We'll talk about a weed called jatropha that grows wild in Africa and has the potential to transform that poverty-stricken continent into the Saudi Arabia of biofuel.

We'll talk about those cities being built each and every month in China, and how Chinese leaders are coming to embrace green policies and practices in order to elevate their nation's standing—and economic status—in the worldwide community.

We'll talk about how investors in even Abu Dhabi, arguably the oil-richest city in the world, are starting to put their money into green energy.

And we'll talk about plastic. Not the plastic we know now as a petro-leum product, but bioplastic, which can be a value-added by-product in the production of sugar-based ethanol fuel for automobiles. In the foreseeable future we could hold in our hands a plastic that provides personal sanitation while we pick up after our dogs and that dissolves, in a brief span of time in a landfill, back into sugar. When that happens, even Liberty, our well-loved but essentially color-blind terrier, will see in green.

1

GREEN MONEY

THE FUTURE LIVES ON U STREET

AT FIRST GLANCE, THE NINE-BLOCK STRETCH OF U STREET IN NORTHWEST Washington, D.C., from 9th Street on the east to 18th and Florida Avenue on the west, might look like any other lively urban neighborhood in the process of reinventing itself—young professionals descending from trendy new lofts and condos to hustle to work, clutching cups of take-out coffee, passing hip new shops and restaurants on the street level as they open for the day's business, dodging scaffolding set up around vintage buildings being reclaimed for new uses. But something is going on here that a casual observer can't see. A significant chunk of the street's infectious energy is fueled by clean, renewable wind power. Thanks to the ingenuity of some of its merchants, U Street is on the cutting edge of the world's energy future—and that's fitting, because U Street has always been a little bit ahead of the curve. Its residents and business owners are used to making history.

In the years before World War II, this stretch of U Street was known as Black Broadway. It was a glittering entertainment corridor where artists like Cab Calloway, Pearl Bailey, and Jelly Roll Morton performed in legendary jazz clubs. Much of the neighborhood's rich Victorian architecture was designed by black architects, and in its elegant town houses, prosperous storefronts, and vibrant community centers, it nurtured national and international leaders in science, law, and the arts: Thurgood Marshall, Duke Ellington, Langston Hughes. Some of the nation's first civil rights protests took place on the streets of this important African American cultural center.

Then, in 1948, the Supreme Court handed down *Shelley v. Kraemer*, a decision that struck down racially restrictive real estate covenants and allowed African Americans to live and buy property wherever they chose. Property owners on U Street began to take part in that era's residential phenomenon—the rush to abandon cities for life in the suburbs—and U Street began its decline. In 1968, 14th Street and U was the epicenter of the riots that followed the assassination of

Dr. Martin Luther King Jr. The violence seemed to seal U Street's fate as a tract of blighted buildings, wandering homeless, and drug traffic so infamous even the police avoided it.

But in 1991 a new metro stop opened at U Street on the city's fittingly named Green Line. Artists and young professionals began to flock back to the neighborhood, lured by cheap rents on the beautiful old town houses and retail spaces. Life—and nightlife—started to return to U Street. These days a person can once again walk down the U Street strip to the sounds of jazz—and hip-hop, punk, and electronica as well. The landmark jazz club Bohemian Caverns, where John Coltrane and Miles Davis once held forth, has added reggae to its repertoire. Busboys and Poets features a performance space where, depending on the night of the week, you can take in a lecture, see a film screening, or listen to a poetry reading or spoken-word event. Tryst is the local wireless watering hole. At Ben's Chili Bowl, one of the few businesses that survived the neighborhood's toughest years to celebrate its current renaissance, you can order up a bowl of the famous beef chili that once made Nat King Cole a regular.

U Street once again glitters. It's a shining example of one of our own era's residential phenomena, the revitalization of our cities. Unfortunately, U Street's business owners are starting to feel the price of their success. Most of the street's merchants don't own the spaces where they do business. Rather, they are tenants. As the neighborhood is once again becoming fashionable, property values have risen—and so have the rents.

In 2006 the nonprofit Latino Economic Development Corporation stepped in and formed a coalition of small, independent businesses, including many on U Street, to help the owners find ways to cope with the pressure of the rising costs of doing business. Ayari de la Rosa, the corporation's business program manager, negotiated group rates for insurance, advertising, and marketing for the coalition members. She also brought the merchants together with Gary Skulnik, a cofounder of

Clean Currents, a two-year-old renewable power brokerage and consulting firm based in Rockville, Maryland, and the former executive director of the Clean Energy Partnership (CEP), a nonprofit agency that promotes solutions to environmental problems through a variety of avenues, including the use of green energy.

Skulnik arranged a deal with Pepco, the main electricity company in the D.C. area, and Washington Gas Energy Services, Inc., a CEP member and alternative supplier that buys energy from wind farms in West Virginia, Pennsylvania, and other mid-Atlantic states, for the U Street merchants to buy "wind certificates" that effectively switched the source of their electricity to 100 percent renewable wind power.

Wind certificates, more formally known as renewable energy certificates, or RECs, are issued to cover a combination of renewable energy sources including wind, solar, and geothermal power. We'll get into how RECs work—and how they might work better—in more detail in chapter 7. For now, let's focus on the benefits of the RECs to each entity involved in the deal Skulnik brokered for the U Street merchants.

Pepco has been recognized by the U.S. Environmental Protection Agency (EPA) as the fifth largest purchaser of green power in the country. In addition to supplying over 110 million kilowatt-hours of renewable energy certificates annually to the EPA's headquarters in Washington, it is also a partner in that agency's Green Power Partnership, a voluntary program working to establish the purchase of green power as a standard for the best practices for environmental management. The Green Power Partnership currently includes over 550 members—Fortune 500 companies, the U.S. Air Force, the World Bank, states, universities, school districts, and large retailers, including Staples and Whole Foods Market. In extending the ability to participate in using renewable energy sources to small businesses, Pepco is broadening its base as an industry leader in providing renewable energy.

The environmental benefit of the U Street deal, according to Skulnik, is that the equivalent of about 2.8 million pounds of CO_2—carbon dioxide, the main greenhouse gas contributing to global warming—is being removed from the air. The carbon offset is equal to taking 185 cars off the roads.

The direct benefits to the U Street merchants include a seamless transition from conventional power. Their group purchase of nearly 2 million kilowatt-hours of wind energy a year has required no capital outlay for or installation of special equipment and no changes in the ways they do business. They are even still billed through Pepco. What did change was the amount due on the bottom line of their electric bills. The U Street merchants expect to save about $21,000 collectively in their first wind-powered year, or an average of 9 percent each. And the savings are expected to increase to 12 percent annually in each subsequent year of the three-year deal.

Nazim Ali, whose parents, Virginia and Ben, founded Ben's Chili Bowl in 1958, is quick to point out that no one in the merchant group entered into the wind power program thinking that they were really going to save money. "We didn't do it to be trendy either," he told us. "It was just the right thing to do for the environment and, it turned out, it made economic sense for all of us too."

As individual businesses, buying wind power might not have made much sense for the U Street merchants. Blending renewable energy into the mix of the power sources one's electric company uses and delivers to one's door is an abstract, and perhaps even daunting, concept for most consumers. In the past, renewable energy was indeed thought of as restricted to the major players—large electric users, such as universities, hospitals, hotels, and government offices. But by pooling power usage in order to buy in volume, and with brokerage firms like Clean Currents to build bridges between power companies and power consumers, it's possible for small business owners and even home owners, as groups, to start reaping the economic benefits of clean, green energy too.

And that's exactly what may be surprising: There are economic benefits in being green.

Our world community is linked as never before by communications technologies that were, as little as twenty years ago, more science fiction than fact. The ability to communicate instantly, over long distances, with people in once-remote corners of our world has enabled us to do business—producing ideas, goods, and services—in ways that we hadn't even imagined just a generation ago. Journalist and best-selling author Thomas Friedman coined the phrase "the flattening of the world" to capture the way this technological leveling of the playing field has sparked development and competition to create a truly global marketplace.

But the ideas we exchange over the worldwide network of fiber optic cables that connect our Internet services and iPhones haven't only to do with commerce, of course. Cultural boundaries that used to isolate us as effectively as geographical ones begin to dissolve as we connect with people and customs we once thought of as so very differ-ent from our own. Through the Internet, for example, I can access almost instantly a recipe for the preserved lemons I liked so well when they were served to me last night at my neighborhood Moroccan restaurant on New York's Upper East Side. I can also log on to www.kiva.org, a Web site that's a conduit for empowering third world entrepreneurs, and make a quick fifty-dollar loan to a businessman in Kenya who needs cash to expand the stock in the small clothing store that supports him and his family. Technology allows us to more abun-dantly appreciate the fruits of other cultures, and it allows us to be more keenly attuned to—and personally involved in alleviating—hardship in parts of the world that might not otherwise have made a blip on our radar screens. As we become ever more tightly interwoven as a global village, our neighbor's welfare increasingly intersects with our own. We become more deeply aware of just how consistently the people of this world dream in common. Economic security for our families, education for our children, streets that are safe to live on—no matter our country, ideology, or income, these are core concerns that can keep most of us up at night.

But there is no concern we people of this world can share more in common than for the greening of the small blue planet we all call home.

Think about it for a minute. We are surely living in a thrilling moment in time—when technology enables us to unite in a cause that we have in common with every other person living in our entire world. It would be difficult to overstate the magnitude of this phenomenon: At no other time in history have so many people, from so many diverse places and walks of life, been able to come together for the same purpose—or been motivated to do so. The challenge of rethinking and reorganizing how we create and use energy is huge, but it's a challenge the whole world can work on. Together. We have the opportunity, right now, to make a green, renewable future our greatest common dream come true.

So, just how do we go about it? How do we translate that lovely but abstract ideal of living our everyday lives with environmental integrity into our reality? Can those of us in the developed world do it while still maintaining our current, comfortable quality of life? Can the great engines now in place in the developing world, India and Asia, churning out what is fast becoming a global middle class, continue their fevered progress at the same time as they reduce their current levels of pollution? What about Africa? Is it either realistic or compassionate to ask people whose urgent, daily concern is for their family's daily portion of food to factor into the struggle what must seem like a luxury—a standard of living that so takes for granted the abundance of its resources that its people can plan ahead for their long-term use?

What is it that will allow sustainable use of our resources to make sense for all of us?

The motivation to live a more sustainable lifestyle likely comes to every individual through a combination of factors. He's frustrated with the high price of gasoline. She's angry at being dependent on a regime that abets terrorism for the daily necessity of oil to heat her home. The environmentalist fears doing such harm to our planet, through the release of carbon gases, that his children will be unable to enjoy the same quality of life as he does now, and the preacher feels the duty to act on the biblical injunction to steward the earth with care.

We come to the idea of saving the earth's resources from many different perspectives, but rarely do any of us list outright, among the benefits of sustainability, "saving money." On the contrary, sometimes even those people who are most environmentally alert go at the problem as if it's a big expensive chore, like putting a new roof on your house or buying new tires for your car, something that simply has to be done no matter the inconvenience or the cost. The reality, though, is that the new roof will increase your home's energy efficiency and likely lower your heating costs, and the new tires will extend your car's gas mileage so you'll have to fill up the tank less often. We need to change the language we use when we talk about environmentalism and renewable energy to include economy; it's a tangible benefit if—as with a new roof or new tires—not always an immediately apparent one, and it's a powerful incentive. A story about the ubiquitous little plastic water bottle can help us to understand just the kind of powerful incentive we're talking about.

Plastic water bottles are made of polyethylene teraphthalate (PET). Like most plastics, PET is ultimately derived from petroleum hydrocarbons through a manufacturing process that emits carbon dioxide, a greenhouse gas. For this reason, plastic water bottles are fomenting a controversy as some people question the need for bottled water at all, especially in the developed world where clean water is abundant. After all, most bottled water is simply filtered tap water with, perhaps, a flavoring added. It's the same product we could produce easily enough for ourselves by installing a filter on our kitchen faucet. But let's try to find the balance in a dilemma that probably almost all of us have experienced: a powerful thirst, the convenience of the 24/7 market, and our common desire to be environmentally responsible.

The average one-liter plastic water bottle that weighs in at 16 grams of PET requires .00052 barrels of oil to produce. Sales of bottled water in the United States far outstrip those in other parts of the world—according to the Beverage Marketing Corporation, Americans alone consumed 31.2 billion liters of bottled water in 2006. Extending the calculation, then, we can know that approximately 16.224 million barrels of oil were used to make the plastic containers for America's water. That's about three-quarters of one day's consumption of oil in the United States.

The National Association for PET Container Resources indicates that a total of 1,170 billion pounds of plastic bottles—water bottles, soda bottles, milk bottles—were recycled in America in 2005. When the bottles are recycled, they are first delivered to a material recovery facility (MRF) for sorting, and then the sorted plastics are baled to reduce shipping costs to reclaimers. The process of reclaiming plastic involves chopping it into flakes, washing it to remove contaminants, and reselling it to an end user that will manufacture new products with it: clothing, carpet, or more PET bottles. This is a classic cradle-to-cradle scenario.

But America lags far behind many nations, especially those in Europe, in its recycling efforts. What happened to the 3,905 billion pounds of plastic bottles Americans bought but didn't recycle in 2005? A lot of the PET containers ended up in the ocean, where the plastic breaks down into tiny but durable particles and is ingested by open-ocean filter feeders—sea life such as sponges and corals, and whales—and eventually kills them. Some of the bottles end up in landfills, where, like the plastic bags we talked about in our introduction, they may never completely decompose. Others are converted to heat and electricity in a waste-to-energy (WtE) process, usually through incineration, which, in turn, creates more carbon emissions. In every case, throwing a plastic bottle in the garbage can rather than in the recycling bin further degrades the environment in some measurable way.

Environmentalists were outraged at the impact the plastic was having in the oceans and landfills and on carbons emissions. They pressed their municipalities to enact more exacting recycling laws—and eleven states, including New York and California, now have deposit laws. They put effort into public education so more consumers would recycle their plastic bottles. And they put pressure on the beverage makers to use less PET in the containers they put their products in. The PET bottle problem was being addressed by logical *environmental* actions.

But let's look at the problem from another angle. Remember the .00052 barrels of oil it takes to make each PET container? When you multiply that by the 3,905 billion pounds of plastic bottles that weren't recycled in 2005, you come up with the remarkable fact that Americans threw 95.735 million barrels of oil in the garbage that year. And the water, soda, and milk bottlers paid for it.

That's why, in the last several years, the average plastic water bottle has been redesigned to weigh in at just about 16 grams, down 5 grams from its former weight.

It's also why bottlers are now embracing the recycling measures they had in the past consistently opposed and are beginning to make their new bottles from reclaimed PET. They found that by using less PET in the first place and by reclaiming recycled PET, they were paying for less of the raw material they needed to make the bottles, oil. And paying for less oil is having a positive impact on their bottom lines.

Right now, the bottled water industry is making its plastic bottles from, on average, about only 10 percent reclaimed PET. But what if laws were put into place that required the companies to use 20, 50, or even 100 percent reclaimed PET in their bottles? Think of the savings they could achieve! It could be the ultimate win-win-win situation: The environment is protected, the corporations show more profit, and the consumer gets to have a convenient and guilt-free long, tall drink of bottled water.

Perhaps the most compelling reason of all to get serious about sustainability and renewable energy, though, is consideration of our future energy needs. The problem should be sobering to even the most enthusiastic Hummer owner. The numbers can seem staggering. That's why we suggest you look at them from the perspective of a homeowner preparing her annual household budget. She knows what her monthly expenses are going to be—her mortgage, utility bills, credit card payments—and she can calculate to a fair degree what her seasonal expenses will be too: She knows that she needs to budget more for heating bills in the cold months of January and February, more for education in the spring and fall when her son's college tuition comes due, and some amount for gifts in December, during the holidays when she knows she'll want to buy presents for friends and family. In other words, she prudently takes into account her future needs when deciding how best to allocate her assets.

The challenge of meeting our future global energy needs is more complicated than putting together your standard household budget, certainly. But if we consider it within the framework of how best to budget our assets—how best to direct the human energy and talents and the funds we have now at our disposal—we can start to break down the sobering, staggering numbers in ways that are similarly simple to conceptualize.

According to Ernest J. Moniz, a professor of physics at the Massachusetts Institute of Technology and a former under secretary of the Department of Energy, world oil consumption is forecast to increase by 60 percent over the period from 1999 to 2020: from 75 million barrels a day to 120 million barrels a day.

In the same period, global demand for electricity is predicted to grow by three-quarters, with the demand in developing Asia alone expected to grow by 150 percent.

Using these forecasts, experts can tell us that 60 percent of the facilities we'll need in a mere twenty years to meet our energy needs are not yet even under construction.

Where are we going to get this energy? Some of the world's oil supplies are, in fact, running out. Others are located in places that are controlled by hostile regimes. Still other deposits of fossil fuel are located in lands so distant from population masses where the energy is needed that transporting it to the end user becomes problematic.

Even if world oil supplies *could* keep pace with the rate that demand for energy is growing, would we want to remain as dependent on oil as we currently are? Given that science has provided us with fair warning about what carbon emissions are doing to the earth, would we want to continue—to *increase*—the risk to our planet?

How long can we wait to start calculating our future energy budget? Consider these stories. South America just experienced its coldest winter in decades. Argentina was forced to cut almost 90 percent of the natural gas it normally exports to neighboring Chile in order to stockpile it for its own emergency needs. Chile was forced to switch to diesel and fuel oil to meet its energy demand—and as a result, its costs of producing electricity nearly quadrupled, and its cities reported the highest number of dangerous smog days yet in this century.

Now let's go to the United States. Power plants ideally function at a 15 percent reserve margin. This means that 15 percent more production capacity is available than is needed on a peak day. As demand for energy grows, power plants can use their reserve margins to temporarily meet an immediate increase in customer usage until new plants are built or an existing one is expanded to accommodate the higher demand. During California's energy emergency in 2000, reserve margins at the plants supplying the state with power shrank to 1 percent. Today, depending on the market area, some reserves are going as low as between 5 and 9 percent. As it takes an average of six years to build just one new power facility and make it operational, it is clear that in parts of the world, the energy future is already a ticking time bomb.

Before we can begin to answer the questions we've raised here, we need to add one more variable to our future energy equation. Our current power distribution infrastructure—the grid that delivers electricity to our homes and businesses—is inadequate to meet the improved power quality we require in our increasingly digital world. In case anyone thinks that the bulk of our energy predicament is focused on developing nations and that, therefore, the problem is really more of a local one for those regions to manage if they can, we have to remember that the developed world is more plugged in than ever before. We microwave; blow dry; air condition; heat; brush our teeth; charge our family's cordless and cell phones; entertain ourselves with TiVo and video games; use our computers to run businesses, help our children with their homework, find recipes, do our banking, pay our bills, purchase groceries, read the news, get directions to the new restaurant everyone says we should try—and we expect that computer to respond *fast* when we need to find out where and how far we have to drive to go eat dinner. We are *hungry*! The same powerful technology that allows us to unite globally in common concern for our world's environment and energy needs feeds more electricity into our homes and businesses than we have ever before required of a system that was designed in the early decades of the twentieth century. And our demand escalates daily. How long can we wait to redesign and build an infrastructure that will efficiently support our modern needs?

The answer, of course, is that we can't wait.

Here's one more fact to keep in the back of your mind as you read and ponder with us the best solutions to our global energy challenge. In 1950, according to the U.S. Census Bureau, the world's population was 2,556,518,868; today it is 6,602,274,812. Today there are well over twice as many people in the world as there were in the decade when the U.S. Supreme Court banned school segregation, Fidel Castro came to power in Cuba, and Nikita Khrushchev banged his shoe on the table at the United Nations. In the years since the first Xerox machine and the first self-service elevator drew power from the electric grid and work on the U.S. interstate highway system first began, we've more than doubled the number of people on the planet, and every one of us wants to plug in and gas up.

In order to keep pace with the voracious needs of our world's population, we have got to diversify our sources of power and fuel to augment traditional fossil fuel sources. We have got to invest in research and development of renewable energy technologies and in the infrastructures that will allow us to seamlessly deliver renewable power and biofuels to the consumer. We have got to increase the efficiency with which we use the energy we already have. And we have got to investigate green technologies that don't yet exist and sponsor development of those that hold promise.

We need to be as prudent as the homeowner with her annual budget, meeting today's needs while planning for the crunches tomorrow certainly has in store.

And we have to be as bold as the U Street merchants, ready to embrace the unaccustomed, and all of its yet-unimagined benefits, as we step into the future together.

2

ENERGY PRESENT

EVOLUTION OF ENERGY

A FRIEND OF OURS TELLS THE STORY OF A FELLOW WHOSE FAMILY ONCE OWNED a small home-heating oil and gasoline distribution business. The business had been providing his family with a fine living for a generation, and when he inherited it in the 1960s, he expected that it would continue to support a comfortable, upper-middle-class lifestyle for him.

Rising oil prices, however, began to cut into profits. More stringent environmental regulations, and internal changes within the oil industry itself, further narrowed the margins. He grew frustrated with the market and angry about having regulations imposed on him that had never hampered his father's way of doing business. He spent a lot of energy— and a lot of the money that the business still generated—supporting politicians who seemed to want to roll back the new environmental standards and loosen up, once again, his business practices.

But as we pointed out in the introduction, that's not the way the collective conscience of humanity has responded to the information it has absorbed in the forty years since the fellow inherited his oil business. Science has presented us with evidence that our reliance on fossil fuels is harming the health of our planet. World events have demonstrated that dependence on foreign oil limits the strategic options of affected nations. And—every bit as important as those two reasons—it has become clear that, in order to keep pace with our galloping global energy demands, we need to augment and diversify our current sources of power. That we've got to wean ourselves from fossil fuels as our sole source of energy. The world is in no position to roll back its thinking about our energy challenge, nor does it give any credible indication at all that it wants to.

By the early 1990s, the small heating oil and gasoline jobber had decided that the limited volume his business was capable of handling under the new government regulations and oil industry standards didn't make it worthwhile for him to continue to navigate what to him was a confusing new set of business practices. He sold what had deteriorated into a marginal business to a larger outfit.

But that's not where this story ends. The small distributor still retained ownership of the valuable parcels of land where his distribution centers and gas stations had been located. Our friend suggested to him that it would be fitting—almost poetic—if some of these sites were transformed into sales offices and warehousing space for solar energy panels—if the one-time oil jobber set himself up as a distributor for a new kind of energy of the future. This was still in the early 1990s, when the idea of installing solar panels in private homes and businesses was just dipping its toe into the mainstream as a viable way to augment power. But our friend reasoned that, as acceptance of solar energy panels continued to grow and become more of a standard practice in building and remodeling, the distributor would be in an enviable position, poised to supply a loyal and long-standing customer base with energy, albeit in a distinctly new format.

But it's very difficult for human beings to release that to which we are accustomed, that which has been reliable and convenient and seemingly seamless in our lives, and reorient ourselves. The distributor's response to our friend's suggestion sums up succinctly the most extreme example of the reflexive reaction to stepping up and facing the limitations of our traditional sources of power: *Are you crazy?* In spite of having spent almost twenty years riding the roller coaster that was a small oil distribution business in the last quarter of the twentieth century, the distributor thoroughly disparaged the idea that any source of power would ever replace good old fossil fuel.

More than developing confidence in the reliability of new energy technologies, and even more of a challenge than the admittedly complex task of designing and constructing an infrastructure that will allow the consumer access to them, what stands in the way of our apparent will to wean ourselves from conventional energy sources is the reluctance to wean ourselves from old ways of thinking.

Before we talk in any more depth about our energy future, reviewing our energy present might help us to move more nimbly beyond that sticking point. How does the world use power and fuel right now?

Where do we get it, and how did we become such rapacious consumers of it? How did our seemingly insatiable appetite for energy *evolve*?

Energy is defined as the capacity for work, and our world's fundamental source of it is the sun. The sun gave our most distant ancestors light and heat, and it is sunlight that plants and phytoplankton convert, through the mystery of photosynthesis, into stored energy. The sun is still the foundation of our energy system; after all is said and done, whenever we burn a fossil fuel, what we are really doing is converting the stored solar energy of prehistoric vegetation into electricity or gasoline.

Living creatures augmented the sun's energy with muscle power, and early humans harvested two other main sources of power: animal power and, upon discovering how to make wood burn, fire. Fire, of course, altered profoundly the way early humans could live on the earth. It expanded the places our ancestors could explore by allowing them to transport heat into colder climates, and it exploded the range of foods they could eat as the process of cooking made previously inedible plants and animals tastier, more digestible, and, in many cases, more nourishing.

Fire was our first evolutionary energy leap. It's tempting to stop here and keep the history simple by assuming that for the first several millennia of human existence, muscle power, animal power, and fire weren't merely the three major sources of energy but the only sources. That's just not the case. Archeologists have found evidence that today's three major sources of energy—the fossil fuels coal, petroleum, and natural gas—were being used by humans as early as 3000 BCE.

In 200 BCE, for instance, the Chinese were putting natural gas to work in gas-fired evaporators rigged with bamboo pipes that they used to make salt from brine. In the seventh century BCE, Persian chemists discovered that by mixing petroleum with quicklime they could make

"Greek fire," a weapon that was very much the napalm of its day. Coal, scientists believe, was used sporadically by cave people, as an innovation by the Romans in England in 100 CE, and routinely by Native Americans for cooking and heating, and to bake clay pottery. By the 1400s, with the invention of the firebrick in England, coal quickly became the home heating fuel of choice, as it burned hotter and more efficiently than wood which, until that time, had indeed been the world's primary fuel.

During the Industrial Revolution of the 1800s, however, humans began to harvest fossil fuels in earnest. And when we began mining and drilling and siphoning to drive the great economic engines of the new industrial age, we went at it as if there were no tomorrow. The use of coal as a fuel rose sharply in the first half of the nineteenth century, with the invention of the steam engine. Steamships and steam-driven locomotives were the main forms of that era's mass transit, and they both used coal to fuel their boilers. Oil came to prominence with the invention of the internal combustion engine and the discovery of petroleum in an experimental drilling operation in Titusville, Pennsylvania, in 1859. Natural gas was used mainly for lighting from about 1890, but in 1925, when a material for pipes was found that made it possible to transport natural gas great distances and in significant quantities, it quickly took its own place in the modern energy mix.

The point is that every time a new discovery was made or an invention was created that made manufacturing more cost-effective, or travel speedier, or life a little more leisurely, the market moved to adapt itself—to *evolve*—to take advantage of each new innovation and the fuel that served it best. Old technologies were replaced and set aside in favor of continuing progress and economic expansion.

The market can continue to do this today. And in response to contemporary environmental concerns, national security issues, and the massive global growth that even today strains our capacity to produce energy, that's what the market has got to do.

As an example of how this can be done, let's focus for a few minutes on what was probably the last great and *conscious* evolutionary step taken to energize the now-developed world: electrification. To keep our discussion of electrification compact, we'll focus on how the process worked in the United States.

In the early decades of the twentieth century, only about 34 percent of the population of the United States had access to electricity. These plugged-in people were, in the main, city dwellers and factory owners who had recently replaced large and cumbersome manufacturing equipment powered by coal-fired steam engines with sleek new machines driven by electricity. Most of America—its small towns and villages and its farms in the heartland—remained in the dark, and stuck in drudgery. Though there were over thirty state or rural power initiatives during the 1920s and early 1930s, they failed time and again to deliver affordable electricity to these remote locales. President Herbert Hoover refused to step in to light America's way, arguing that bringing electricity to all Americans was not the responsibility of the federal government. In contrast, Franklin Delano Roosevelt, then governor of New York, aggressively promoted his state's electrification and in 1931 created the New York Power Authority to develop an inexpensive new source of hydroelectric-generating capacity along the St. Lawrence River.

By the time Roosevelt took office as president of the United States in March 1933, the country was mired in the Great Depression. Many rural and state power authorities had collapsed, and private funding sources had backed off from investment in a rural power infrastructure. The cost, the investors believed, was too high—and the potential profits too low—to justify electrifying areas where there was a low population density.

But Roosevelt still believed passionately in turning on the lights for all of America. Without city conveniences like electric lighting, or the ability to turn on a radio and so participate in the news of the day, or the means to automate equipment and take the backbreaking nineteenth-century drudgery out of farm life, Roosevelt believed that American citizens were disadvantaged for life in the twentieth. That America itself would be at a disadvantage to compete in a marketplace that had been expanding steadily, since the end of World War I, outside of its own borders. The path to prosperity, Roosevelt maintained, was necessarily dotted with electrical poles.

In May 1935, Roosevelt put into place an ambitious plan by signing Presidential Executive Order 7037, which created the Rural Electrification Administration (REA).

The REA is now widely considered to be one of the most stunning success stories in the history of federal policymaking. Its impact on the national economy was immediate—just two years after its inception, 350 projects in 45 states were delivering electricity to 1.5 million farms. And its impact was profound—it paved the way for the economic boom that would follow in the decades after World War II. In quite literally turning on America's lights, it laid the foundation that enabled the nation to earn a leadership position in the now-global marketplace.

In subsequent chapters we'll refer back to the REA. Portions of its basic structure can serve as a model for the way in which the transition from dependence on traditional power sources can be accomplished today as we stand at the dawn of another energy evolution.

If you still hold any doubt that we are right now pioneering a brand-new era, consider this: According to the U.S. Department of Energy "Many energy experts believe that the age of fossil fuels is only an interlude between pre- and post-industrial eras dominated by the use of renewable energy." Nations that will be in a position to lead in the post–fossil fuel age are the ones that are positioning themselves now to take advantage of the technologies that will keep them in renewable power.

For the moment, however, we are still living in the interlude, the fossil fuel age. The rest this chapter is devoted to a brief look at how we currently use nonrenewable sources of power and fuel. A comprehensive report on all aspects of the status of fossil fuels and our use of them would be a book unto itself; the discussion that follows is an overview of the options available to us as we move from one energy age into the next.

"OLD KING COAL"

"I remember being fascinated, as a child, by the coal bin in my family's home—the heavy metal hatch that the driver would swing open before he backed his big red delivery truck up to the basement, the rattle of the chunks racing down the chute, the piles of coal, taller than either of my parents, in the small room below. I pretended that it was a secret passageway to a hidden room and I loved to play in its darkness—castaway, or secret agent—though I wasn't allowed to play in there; I would get covered with coal dust and my clothes would be ruined and I got in trouble every time," our friend Lynn tells us. "I suppose that's why they finally decided to teach me how to stoke the furnace. Shovel in the coal and shovel out the ash, and empty the ash can after it cooled. It was heavy, filthy, searing hot work, and it had to be done first thing every morning. It was the most effective cure for my obsession with the coal bin that they could have come up with."

As long ago as the beginning of the Middle Ages, coal was supplanting wood as the preferred heating and cooking fuel. As early as 1570, England was producing enough of the "black rock" to support a flourishing export industry including, eventually, great shipments to the American colonies. Though its use to heat homes would taper off, giving way to less messy energy sources—oil, natural gas, and then electricity—as the twentieth century marched on, coal remained the primary home heating fuel throughout the 1950s and into the 1960s. Today coal is the primary fuel used globally to generate electricity, accounting for a full third of the world's electricity production and for 54 percent of America's electricity alone.

Why is coal so popular? For several very logical reasons. First, it's the least expensive of the fossil fuels. It costs about 80 cents to generate a million British thermal units (Btus) of energy using coal as a fuel source. Compare that to the $3.24 per million Btus for natural gas or the $4.61 per million Btus for crude oil.

Coal is so cheap because it's so plentiful. Experts estimate that in the United States alone, there are enough deposits of coal to meet the nation's energy demand—to supply electricity at our *current* rate of use—for another 250 years. In Australia's Latrobe Valley, there is a seam of coal 35 miles long and 600 feet deep—so gigantic that after nearly a century of mining, more than 95 percent of it is still waiting for harvest.

Also, unlike the case with oil or natural gas, we know almost exactly where to find deposits of coal and, with advances in mining techniques over the years, both in strip or "surface" mining and in shaft mining, we've increased the efficiency with which coal can be removed so that we can rely on a steady supply of it.

But coal, as our friend Lynn knows firsthand and all too well, is also filthy. Way back in 1306, people were already complaining of its smoke and noxious fumes to the point that Edward I of England actually tried to have coal-burning fires banned for any use save blacksmithing. King Edward's action, possibly the first-ever attempt at environmental mitigation, was unsuccessful. His subjects might have hated coal's smoke and fumes, but they were not at all disposed to giving up its cozy warmth or its convenience as a cooking fuel.

Seven hundred years later, our perspective is a little different—and we have the science that King Edward lacked to confirm coal's toxicity.

Coal, like all fossil fuels—like, indeed, all living things—is made up of carbon. When it's burned, coal's carbon combines with oxygen in the air to form carbon dioxide (CO_2), the most problematic of the greenhouse gases. Greenhouse gases include water vapor, methane, and ozone as well as carbon dioxide. In naturally-occurring quantities these components of the atmosphere absorb the longer wavelengths from the sun, creating earth's habitable temperatures. And although coal is plentiful, there's no getting around the fact that it's also our most inefficient fuel in terms of the pollution it generates: Burning just 3 million tons of coal releases a whopping 11 million tons of CO_2 into the atmosphere. Coal-burning power plants are estimated to release 11,357.19 metric tons of carbon into the air annually, or over 40 percent of worldwide carbon emissions.

As if carbon dioxide weren't a big enough problem all on its own, there are also other impurities trapped within coal's carbon, including sulfur, nitrogen, and mercury, that burning releases into the air to create acid rain, or flushes into the seas where it poisons the habitats of sea life. Food fish. Like tuna.

The United States' big, churning coal-fired power plants are responsible for a lion's share of the world's coal-produced pollution. For example, over 65 million tons of carbon dioxide were emitted in 2005 in the Washington, D.C., area alone, more than all of the emissions from Hungary, Finland, Sweden, Denmark, and Switzerland combined, thanks to the three coal-burning power plants that call the U.S. capital area home. The United States has taken the lead in trying to find ways to mitigate the problems; in 1985 the U.S. Department of Energy instituted the Clean Coal Technology Program.

In the over twenty years that have passed since then, the debate about whether there really is, or can be, such a thing as "clean coal" hasn't been settled. Reasonable people can disagree as to the fact or fiction of clean coal—indeed, Michael and I have very different opinions about it; I think "clean coal" is an oxymoron.

But we have to approach the pollution inherent in burning coal as a fuel from a very practical point of view. As we noted a few paragraphs ago, coal-fired power plants account for a third of all of the electricity generated worldwide. No matter how problematic the pollution generated by coal, an immediate, abrupt stoppage in its use is impossible. Without alternative energy sources in place to take up the slack from the abandoned coal-fired plants, it isn't only global economic growth that would grind to a screeching halt; the routines of our lives would be profoundly disrupted. Our lights would go out. Our hot water heaters would go cold. The computers and fax machines and other appliances we use in our homes and businesses to enable us to function day to day would fade to black. We'd be out of the game entirely.

It seems likely, therefore, that coal, whatever its drawbacks—and there are plenty of them—will remain a prominent player in the world's energy mix for some time. But that's not a wholly tragic scenario. Because coal is cheap and plentiful, its continued use offers us the time—the wiggle room—we need to put into place renewable power generation systems and to build the infrastructure to support these renewable systems. Depending on coal as more than a stopgap measure would be foolhardy, but the fact that we can use it to give ourselves the breathing space we need to transition gradually and wisely into a clean energy future can be seen as the silver lining in coal's dark gray cloud.

In the short term, so that the air in our temporary breathing space is as pure and healthy as it can be, a prudent option is to take advantage of the continued development of the technologies that can remove some of the worst pollutants produced by burning coal.

CLEAN COAL

Cleaning up coal is a lot like trying to keep your house in order in the middle of a two-year-old's birthday party. There are so many inherent potential disasters in a room full of toddlers. Before the little guests arrive you baby-proof your house, but one of them inevitably finds a way to get a grip on the tablecloth and topple the silverware and napkins. The guest of honor opens his presents and the gift-giver starts to wail because she wants the present for herself. You serve the cake and ice cream and, just when you think the kids are content with chocolate, you turn around and one of them is smearing it on the wall. A game of tag ends with tears and a scraped knee, a popped balloon is cause for hysteria, and one tiny attendee is scared to death of the clown you hired to entertain.

Now imagine that every child at this party is just beginning his or her potty training.

Trying to capture and control all of coal's inherent potential disasters is a damned tricky business.

Sulfur

Sulfur is one of the causes of acid rain—any precipitation that has a high, man-made acid content and is thus harmful to plants, animals, aquatic environments, and buildings—and there are two sorts of it in coal. Pyritic sulfur—so named because it's combined with iron to form iron pyrite, also known as fool's gold—can simply be washed out of the coal. In coal preparation plants, the coal is crushed into small chunks and fed into a large tank of water where the coal floats and the sulfur sinks.

The other sort of sulfur—called organic sulfur because it's chemically connected to the coal's carbon molecules—is more difficult to deal with. Scientists have tried many ways of removing, or at least of reducing, the content of organic sulfur in coal, but so far they've all proved to be quite expensive. The most feasible solution to date is a contraption known as a flue gas desulfurization unit. These units are commonly called scrubbers because their job is to "scrub" the sulfur content out of the smoke released from boilers after the coal has been burned. Scrubbers work by spraying water mixed with powdered limestone into the flue gases. The limestone acts as a sponge, pulling the sulfur out of the gases and leaving behind a wet, white paste or, in newer units, a dry, white powder—a substance known in either case as sludge.

Nitrogen

About 80 percent of the air we breathe every day is made up of the element nitrogen, so, on its own, nitrogen is benign. When it's mixed with coal, however, and burned at the high temperatures necessary to turn coal into electricity, its atoms join with oxygen to form nitrogen oxides, which cause smog and turn the air a brownish color. The most common and cost-effective way that science has devised for reducing the nitrogen oxides that result when coal is burned is to try to keep the nitrogen from mixing with the oxygen in the first place. This is done through a process called staged combustion because the coal is burned in stages, in burners where there is more fuel than air in the hottest combustion chambers so that most of the air combines with the fuel rather than with the nitrogen.

Mercury

Coal-burning power plants are the largest human-caused source of mercury emission into the air. The free mercury eventually settles into bodies of water or onto land, where it's ultimately flushed into the seas. Microorganisms in the water convert the mercury into methylmercury, a highly toxic form of mercury that builds up in fish and shellfish and can harm the brain, heart, kidneys, lungs, and immune systems of the people who eat the fish.

The good news, even according to sources like the National Wildlife Federation, is that technologies already exist that can capture up to 90 percent of mercury emissions.

But these clean coal technologies can work only if they're actually installed in the coal-burning power plants. Even in the United States, where the effort to reduce coal-related pollution began more than twenty years ago, 68 percent of coal-burning power plants are more than thirty years old, and many of them aren't yet fitted with state-of-the-science equipment.

Why, if the capability exists to make burning coal a more environmentally friendly practice, isn't the equipment that helps to make it so an industry standard? There are several reasons, but the plainest is that regulations that require the environmentally friendly practices aren't supported by the governments that made the laws. To illustrate this point, let's look at another environmental problem that results from coal usage through the practice of strip mining.

About 60 percent of coal is acquired through strip mining, a mining method that's often referred to as "surface mining" to pretty it up. Gigantic, earth-moving machines remove up to 200 feet of covering soil to reach deposits of coal underground. When the coal has been removed, the land is left severely scarred. Most places where strip mining takes place now have laws that require the land to be restored—recontoured and revegetated—when the mining operation there is complete. In the United States, in the Wyoming–Montana region where the country's largest coal reserves are located, strip mine operators must post a bond as evidence of their intent to reclaim the land after it's been mined. But it is said that in Montana, no strip mine operator has ever gotten its reclamation bond back.

This lackadaisical approach to environmental regulations may, however, be beginning to change. In October 2007, a settlement was reached between the American Electric Power Company (AEP), the largest electricity producer in the United States, based in Columbus, Ohio, and the Environmental Protection Agency (EPA), a dozen environmental groups, and eight northeastern states that had brought a lawsuit against it for failure to install pollution controls as required under the Clean Air Act.

Under the terms of the settlement, the AEP has agreed to clean up forty-six of its coal-fired operations by installing scrubbers and other pollution-control devices to cut emissions of nitrogen oxides and sulfur dioxide, which have caused acid rain so intense over the last quarter of a century that it has eaten away at the Statue of Liberty.

Environmental activists might say that the effort to bring the AEP into compliance with government regulations has been an exhausting one—the lawsuit was originally brought in 1999, and under the terms of the settlement, the company isn't required to get its pollution-control act together until 2018—but let's approach the effort from a slightly different perspective.

Think back to the small oil jobber we talked about at the beginning of this chapter. He grew angry about government regulations that imposed new and expensive environmental standards on the way in which he could do business, and his anger was not wholly unjustified. For a generation he and his family had run their heating oil and gasoline distribution business honorably, according to all of the rules that were in place at the time. But as scientific knowledge advanced, theretofore unknown environmental hazards in the ways that oil and gasoline had been handled began to emerge; in response to this knowledge, government—honorably motivated to mitigate the hazards—changed the rules.

Now, we all know what it's like to play a game honestly and have the rules change on us in the middle. It's exasperating, especially when we'd been winning, and the jobber, as we said, had been working hard to win a good living from his business. What the government didn't do, when it changed the rules about how the jobber could operate his business, was to put into place policies that would have helped him to cope with the new guidelines. It was the costs of cleaning his properties of previously unknown or acceptable levels of contaminants and installing new safety equipment—millions of dollars that the jobber's company simply could not absorb—that drove one small American businessman out of business.

In paving the way for future—and quite possibly *faster*—compliance for businesses as small as the jobber's and as large as a major utility company, we need to couple the creation of new regulations with tax credits and other incentives that enable those businesses to stay afloat as they adapt to the new rules. To continue to do well while doing good. Coal-fired utilities provide us with the essential service of electricity— we can't blame them for doing it profitably; we can incentivize them to do it more cleanly. Among the primary goals of any government—goals that most of us can agree on as fundamental—are protecting the environment, safeguarding the energy supply, and empowering its nation's domestic economies, and frequently one of these goals can't be sufficiently met without catching the others' backs too.

BLACK GOLD. *TEXAS TEA.*

A "seep" is a place in the ground where fossil fuel oil leaks up from the earth below. In the eighteenth century petroleum was in great demand as a fuel for lamps because the price of the more traditional lighting fuel of the time, whale oil, had skyrocketed. Most of the oil at this time was collected from seeps, or distilled from coal into liquid, or by skimming it off the top of lakes.

Then in 1859, Edwin L. Drake, a railroad conductor on sick leave, struck liquid oil at his well in Titusville, Pennsylvania, and devised a way to pump the stuff to the surface. The world has not been quite the same since. Today, 37 percent of the world's energy demand is met by petroleum, and oil is one of the most valuable—and probably the single most volatile—commodities on the planet.

Oil, the 75 million barrels of it that are used everyday somewhere around the world, is pumped from wells located in nearly every country on the globe where a viable source exists. It is shipped by barge and pipeline, train and ship to refineries where the thick, black crude is distilled into the fuels and by-products that have become a part of almost everything we touch, nearly every hour of our lives.

Thirty-three percent of the world's oil is consumed by industry to make the household products we can't seem to live without. To get a better idea of how dependent we are on this physically—and politically—sticky stuff, run a little experiment of the kind I did to figure out just how very much oil is part of the course of your unconscious daily routine.

The first thing I picked up in the morning was my toothbrush, which is made of petroleum-based plastic. It was sprinkling outside on the day that I ran my experiment, and when Michael brought the daily papers in from the stoop, they were sheathed in plastic to protect them. Parts of the coffeemaker were molded from plastic, the interior of our daughter's lunchbox was formed from plastic, and the recycling bin where Michael and I deposited the sections of the morning papers as we finished reading them was itself a sort of monument in bright green plastic.

In the bathroom, my hair conditioner and deodorant both contained petroleum products. The reason my lipstick not only colors but moisturizes is oil. The lavender-scented container candle next to the sink burns evenly and without smoking because oil is part of its manufacturing formula.

Though our home isn't heated by oil—just 11 percent of the world's oil consumption is used these days for home heating—it's likely that some of the electricity I accessed when I turned on my molded-plastic computer to check my e-mail was generated at a fossil fuel–fired power plant. And though I do my best to buy organic products, it's also likely that some of the foods I served my family for breakfast—the toast or the eggs or the slices of apple—were produced with the help of petroleum-derived fertilizers or pesticides.

Oil was used to make the frames of my eyeglasses, the crayons our daughter was using to color in her school project, the detergent I squirted into the dishwasher, the ink I used to write a thank-you note, the bubble gum our daughter was blowing as she colored—and oil was the reason why the burn ointment I rubbed on my finger, where I scorched it taking last night's roast chicken out of the oven, worked so efficiently to stop the pain and start the healing process.

It wasn't even 9 a.m. yet, and already I had used oil in well over a dozen different ways.

Then I got into the car to drive to the office.

According to the International Energy Agency (IEA), a Paris-based organization that acts as an energy policy advisor to over twenty-six member countries, 58 percent of the oil that's consumed around the world is used for transportation: gasoline for cars, diesel for trucks and ships, and jet fuel. The agency estimates that 97 percent of the transportation sector's daily energy needs are currently met with oil, and it projects that in the next twenty years, transportation will grow to account for a full two-thirds of the world's oil consumption.

The transportation sector also accounts for about 20 percent of the world's carbon emissions. That percentage of pollution will only continue to grow in the years ahead, as more people in developing countries acquire cars—and China is the fastest-growing new car market in the world. That's why we need to take two important steps right away: Require automakers to design cars that give us more bang for our gasoline buck today, and commit to the development of alternative fuels as well as the infrastructure that will conveniently deliver the fuels to tomorrow's drivers.

Although oil gets most of the press when it comes to media coverage of fuel and energy issues, we have an embarrassing history of paying only spotty attention to its problems. When the price of oil goes up, or when a major oil-producing region of the world is embroiled in political instability, we all focus on the oil market. Usage drops and we prove, once again, that in times of necessity, we're really very good at conserving this resource.

But then the price of oil drops. Usage goes back up, and we retreat back to our old gluttonous ways. The consistency of our concern about oil—its availability as well as its environmental impact—follows the oil industry's pattern of supply and shortage to an almost shocking degree.

We have, even more distressingly, grown so used to this century's sharp rise in the price of oil that it no longer seems to prompt us to conservation. Today's record price of over $100 a barrel is a long way from the high of $12 a barrel that alarmed us back in 1973. Part of the reason for our growing complacency can be attributed to an adjustment for inflation—$12 in 1973 is over $58 today. But another, and perhaps larger, part of the reason is that in 1973 oil prices quadrupled as a result of an embargo imposed by several Arab exporting nations on countries that supported Israel in the Yom Kippur War. The Arab nations curtailed production by 5 million barrels a day. The supply of oil was severely restricted.

Today, on the other hand, our oil problem is in many ways not one of supply but of demand. Even if you're still undecided about where you stand on the debate about peak oil—are we running out of the stuff or do we have tons of it left?—the simple fact remains that our consumption of it has grown radically and the costs of getting it out of the ground have risen in tandem with our need. Let's review, briefly, some of the other drawbacks of fuel oil to help inspire us to walk to the store for that quart of milk instead of firing up the car engine.

Burning oil causes CO_2 to be released into the atmosphere—that's a no-brainer, at the top of everyone's list. But calculate into the pollution equation the risks of damage that drilling for oil can cause to fragile ecosystems, or the risks of transporting oil, often aboard a ship and across an ocean, and the environmental impacts of spillage and leaks to sea life and along our coastlines.

Drilling for oil anywhere is often a gamble—according to the United States Department of Energy, an average of only 1 out of 3 exploratory wells turns into a producing well and that statistic is even less, just 1 in 5, for wildcat operations—but the two largest reserves of oil left in the world are located in Saudi Arabia and Venezuela, and neither of those regions enjoys strong political stability.

But here's the kicker. Many experts are telling us that peak oil is indeed very real. Like the joke about the restaurant where the diner is complaining about its bad food—the potatoes are greasy and the eggs are overcooked and the bacon is fatty and *the portions are so small*—fuel oil is very expensive; it contributes to global warming and drilling for it scars the earth; it's often located in unstable regions of the world; and *the amount that we are left with is so very small*: Experts estimate that the remaining reserves of oil in the world are just enough to last us for another thirty-six to forty-five years.

Thirty-six to forty-five years.

That ought to absolutely rivet our attention on the search for alternatives.

There are those, however, whose attention isn't seriously engaged by that projection—who believe we aren't, in fact, running out of the resource, who don't believe we have reached or are approaching peak oil. This may qualify as wishful thinking but it doesn't dilute our point one bit. A very affordable $2,500 car has just been introduced in India, which is going to greatly increase the percentage of drivers worldwide; 96 new regional airports are currently being built in China and every plane that lands at every one of them is going to require fuel. Even if we weren't, indeed, running out of oil, our economies are growing too fast to keep up with the demand we place upon this expensive and polluting source of energy.

THE "CLEAN" FOSSIL FUEL

Of all of the fossil fuels, natural gas gets the best press—and for a lot of good reasons. Let's return to the bus lawsuit I mentioned in the introduction to understand why.

In September 2000, on behalf of the Giuliani administration, I helped to introduce a new bill to regulate New York City sightseeing buses. At the time there was a fleet of about seventy sightseeing buses powered with engines that weren't compliant with the current EPA emissions standards. Those seventy buses were, in fact, emitting pollutants equivalent to 1,750 Metropolitan Transit Authority (MTA) buses! The legislation that we were introducing would require the bus owners to follow the federally mandated emissions standards as a condition of being licensed in New York.

But these buses weren't the only ones operating in the city, of course. New York, like most other large cities, has a complex system of public transit, and buses are a vital part of the system. Since the mid-1990s, the MTA had been successfully integrating buses that are powered by natural gas into its fleet.

At the time that they were being introduced, the new buses were the subject of quite a few urban myths, and the biggest ones had to do, naturally, with their cost. The purchase price of the buses powered by natural gas was, indeed, $25,000 to $50,000 a unit more than for a conventional diesel bus. But because the fuel is less expensive, a natural gas-powered bus typically paid for itself in under three years. The biggest savings of the natural gas-powered buses, however, was in the reduction of pollutants.

Natural gas is the cleanest of all of the fossil fuels. On a chemical level, coal and oil are made up of more complex molecules with higher carbon ratios, and higher nitrogen and sulfur contents, than natural gas. Burning natural gas releases, inherently, less sulfur dioxides and nitrogen oxides into the atmosphere. Additionally, while the combustion of coal and oil releases ash particles, natural gas releases virtually no particulate matter into the air. From the perspectives of reducing long-term operating costs and affecting short-term environmental mitigation, the buses powered by natural gas were a wise choice.

Alas, though natural gas is a cleaner fossil fuel, it is not a *clean* fuel. It's true that while burning coal emits 208,000 pounds of CO_2 for every billion Btus of energy, and oil emits 164,000 pounds of CO_2

for every billion Btus, natural gas emits only 117,000 pounds. But that's still 117,000 pounds of greenhouse gas that the atmosphere of our planet can ill afford.

Additionally, the principal component of natural gas is methane. Methane itself is a potent greenhouse gas. It has the ability to trap heat over twenty times more effectively than carbon dioxide, and methane emissions account for 8.5 percent of the greenhouse gas emission based on global warming potential.

Natural gas is burdened with other problems as well. Transporting it to its end users, for instance, has long been one of them. It wasn't until 1925 that engineers devised pipes made of materials that could safely and efficiently transport the gas for distances and in quantities significant enough to make it practical for general use.

Transportation remains a problem today for natural gas because most sources are located in parts of the world far removed from where people who might use it actually live. Experts refer to these far-flung reserves as "stranded," meaning that getting the fuel from its source to its market is a cumbersome and expensive proposition.

When we calculate the efficiency of any energy source, we have to consider the amount of energy that has to be used to get the fuel to market. How much energy is expended in just getting the natural gas from, say, Trinidad or Tobago, the two primary suppliers of the United States?

Some of the exporting costs of natural gas have been addressed through the processes of producing condensed natural gas (CNG) and liquefied natural gas (LNG). LNG, for example, is made by cooling the gas to a temperature of about $-260°$ F, at which point it becomes liquid. The process of liquefying reduces the volume to about 1/600th of the volume of the gas in its natural state, so much more of it can be shipped at any one time, within any one carrier, significantly reducing the amount of energy needed to get it from one place to the other.

On the receiving end of the shipping process, however, natural gas presents other issues. For one thing, there aren't many terminals where the gas can be off-loaded and "regasified." Import terminals are expensive to build, and communities where industry proposes to build the terminals often object for reasons of safety and aesthetics.

Among the safety concerns, members of the health professions link the use of natural gas to increased diagnoses of asthma, allergies, and other respiratory problems. Dr. William J. Rea of the Environmental Health Center in Dallas, Texas, cites "gas cook stoves, hot water heaters, and furnaces" as important factors in the illnesses of over 47,000 patients. Environmentalists and health advocates object to the construction of new natural gas pipelines and related projects until more studies can prove their safety to the humans who live in the communities where the facilities are located.

Natural gas does have its place as we transition from our current fossil fuel age. But, in spite of its real benefits in reducing emissions of certain pollutants, we can't forget its drawbacks, both known and unknown at this time. Natural gas, for all of the good press it gets, is still a carbon-based fossil fuel.

CARBON SEQUESTRATION

If we come to terms with the necessity of fossil fuels remaining the norm—at least in the short term while we transition into our energy future—what can we do to reduce their greatest threat to our planet's health, the continuing emission of billions of pounds of carbon dioxide into the atmosphere? The IEA estimates, for example, that overall world use of coal is going to increase by about 50 percent in the next twenty years. We have to expect a correlating overall increase in carbon emissions. It's clear we have to do something.

The first and most immediate offensive action we can take is to improve the efficiency of our energy use. This is the area where switching from conventional light bulbs to compact fluorescent ones and using a low-carbon fuel like natural gas can make a dramatic impact, albeit in small and incremental ways. The goal of this first phase isn't to cure the problem but to *stabilize* the amount of carbon in the atmosphere. This is a reachable goal if each one of us takes every step available to him or to her to make sure that, individually, we're contributing minimal amounts of pollution to the carbon problem.

The second phase in addressing the carbon dioxide problem is to *reduce* the amount of carbon accumulating in the atmosphere through carbon sequestration, the process of capturing CO_2 emissions and permanently storing them.

There are two sorts of carbon sequestration. The first is known as terrestrial sequestration, a process of nature that has been going on as long as there have been living things on the planet: trees absorbing carbon dioxide. Trees are the lungs of the earth, breathing in the CO_2 and turning it into wood and leaves.

Unfortunately, too much has been made of the concept of an individual planting a tree as a way of offsetting his or her own carbon contribution. It's a little more complicated than that. Planting a tree is, of course, a positive step. But how big a step it is depends on the specific carbon footprint of each person. Calculating our own carbon footprint involves considering everything from the way we heat our home, to the kind of car we drive, to how the shoes we are wearing were manufactured, to, of course, where we live in the world. On average, in the United States, one person's carbon output is 20.5 tons of CO_2 per year; according to the National Park Service the average tree can absorb about 1.33 tons of carbon every hundred years. You can see that making terrestrial carbon sequestration a major part of the solution to the larger problem of worldwide carbon accumulation is going to take a broader management scheme for our forests, agricultural lands, and wetlands.

Let's focus on forests. Scientists who've studied the amount of carbon absorbed by our forests are beginning to understand that, in a tree's life span, it can absorb carbon dioxide efficiently for only a certain number of years. Young trees have a voracious carbon appetite, sucking up CO_2 during their vigorous growing years. As trees age, the amount of CO_2 they can process levels off. Planting new forests on a grand scale, as China has been doing by the millions of acres since the 1970s to control floods and erosion, has, at the same time, soaked up nearly half a billion tons of CO_2 that would otherwise be floating around earth's atmosphere.

Hand in hand with planting forests, however, is their management. How often should these new forests be thinned, and at what age for each particular species, in order to keep the forest young enough to continue acting as a carbon sink?

Complicating the management of forests is the problem of a tree's overexposure to carbon dioxide. Prior to the Industrial Revolution and the corresponding increase in the burning of fossil fuels, the balance of carbon emissions from burning wood was pretty much kept in check by the breathing of the trees. Today our need for energy throws that balance out of whack: Older trees, when exposed to too much carbon dioxide, will die. Whole tracts of forest in Germany, where forests are revered, are now succumbing to the inability of the trees to deal with an overload of carbon.

Factor in natural decay, deforestation, and damage from fires and weather-related disasters such as hurricanes, and it becomes clear why forest management is seen as one of the new century's top job opportunities. Those who take on the task of researching and managing tracts of nature have a critical role to play in restoring the balance of carbon in earth's atmosphere.

The second sort of carbon sequestration involves the potential for humankind to recapture released CO_2 and store it permanently in geologic formations, including oil and gas reservoirs, unmineable coal seams, or deep beneath the ocean. The theory is that carbon dioxide would be separated and captured from emissions directly at its source—power plants, fuel manufacturers, and other industrial systems—before it is emitted into the atmosphere, and then injected back into pockets of the earth.

President Bush announced in 2002 that "our investment in advanced energy sequestration technologies will provide the breakthroughs we need to dramatically reduce our [greenhouse gas] emission in the longer term." But so far very little money has been directed into its research and development, so it can't be determined if this theory for capturing carbon in the atmosphere and storing it is a workable solution. Within the Office of Fossil Energy (FE), a part of the U.S. Department of

Energy, scientists are working on about sixty projects funded by government, industry, and other international partners to gather data about advanced carbon sequestration programs. But plans for the nation's first clean-coal power plant that would have captured its carbon emissions and injected them underground were shelved in October 2007 due to lack of funds.

Until programs that can make clean coal a real possibility are adequately funded, it behooves all of us to turn off the lights when we're not using them, walk where we can instead of driving the car, and, along the way, plant a tree or two.

ATOMS FOR PEACE

This new power, which had proved itself to be such a terrifying weapon of destruction, is harnessed for the first time for the common good of our community.

—Her Majesty Queen Elizabeth II, Calder Hall, 1956

Before a crowd of thousands that included top scientists and other dignitaries from over forty countries, Queen Elizabeth II pulled the lever that directed the world's first power from a commercial nuclear facility to England's national grid. Within four hours the town of Workingham, fifteen miles up the Cumberland coast from Calder Hall, became the first town in the world to light up with nuclear energy. "Epoch-making" was how Richard Butler, the Lord Privy Seal, described the occasion. He predicted that within the decade, "every new power station being built will be an atomic power station."

Three years earlier, Dwight D. Eisenhower, the president of the United States, had made an impassioned address to the General Assembly of the United Nations. The fearful power of atomic energy, he insisted, "must be put into the hands of those who will know how to strip its military casing and adapt it to the arts of peace." Eisenhower called for experts to be mobilized to apply the technology of atomic energy to "the needs of agriculture, medicine and other peaceful activities. A special purpose would be to provide abundant electrical energy in the power-starved areas of the world." He made this speech amid a growing wave of awareness among the people of the developed world that they might be becoming a little too dependent on imported oil, and with the intent to "seek peaceful opportunity for [the people of the East] to develop their natural resources and to elevate their lot."

Now he witnessed, with the rest of the world, Workingham lighting up, the first fruit of the global challenge to put atomic energy to work for a constructive purpose.

In the aftermath of nuclear accidents at Three Mile Island in 1979 and Chernobyl in 1986, the public's warm enthusiasm for nuclear energy in 1953 might well seem chillingly innocent. Of all of the problems imposed by the use of fossil fuels—all of the resultant dangers to the environment or to the public's health—none of them has been as demonized, or produced an outcry of "Not in my backyard" quite as resonant, as nuclear energy.

But 439 nuclear reactors are working right now around the globe, producing 16 percent of the world's electricity. The conventional wisdom might be that nuclear energy was but a detour along our path of energy evolution and that it will pass into extinction as the nuclear facilities still in operation age and become too costly to repair and upgrade. There are energy experts around the world, however, who will tell you that in order to meet growing global energy needs, we are going to have to factor the continued use of nuclear reactors into the mix. If, as the IEA predicts, 27 percent of the nuclear plants now in operation will be retired in just twelve years, that certainly does leave us with a significant energy shortfall—a slack that will have to be taken up by new renewable energy technologies, increased calls for fossil fuels that we can ill afford, or more nuclear power. Perhaps, these same experts tell us, we will need to make use of all three options.

Nuclear energy fell into disfavor in the 1970s partly in response to the accident at Three Mile Island and then later at Chernobyl. This disfavor was fomented partly by citizen protests, and is a result of the terrible, emotional issues of radiation and how to dispose of nuclear waste. Mostly, however, it's a response to the economics of building and maintaining a nuclear site.

Back in 1956, nuclear energy promised, in the popular vernacular, to deliver to people's homes electricity that was "too cheap to meter." The anticipated economies of scale didn't pan out, however.

The truth is that nuclear reactors have always been expensive to build, even in the 1960s and 1970s, amid the world's frenzied development of the new technology. While the infant industry was growing up, it was also learning the lessons of experience. Design changes that arose from adapting the accumulating knowledge of how to operate a nuclear power plant drove up the costs of construction. Safety issues became an urgent concern following Three Mile Island, and the retrofitting that became standard further increased the costs. Shortages of uranium to fuel the reactors also impacted negatively on the bottom line. The public might have rejected nuclear energy because of health and environmental concerns, but, more to the point, industry rejected it because the potential for profit didn't come to pass.

In September 2007, the four giant cooling towers at Calder Hall were leveled with explosives in a demolition that took three years to plan.

The world's first commercial nuclear facility was reduced to a pile of rubble that, it was estimated, would take over three months to haul away. The first monument to the once-so-promising atomic age was a dinosaur, no longer useful, and it was blown away.

A month later, in October 2007, the U.S. Department of Energy announced plans to subsidize the construction of the first nuclear power plants ordered since 1978. Energy secretary Samuel Bodman said in the announcement that nuclear power is "the only mature, emissions-free technology that can supply the power America will need to meet the projected increase in demand for electricity over the next twenty-five years," and he promised to seek even more funding for new nuclear power in the 2009 federal budget.

Then, in November, Areva, the French nuclear power giant, signed the largest deal in the industry's history to bring nuclear technology to energy-hungry China.

Whether this news flooded you with relief, knowing that some step had been taken to address our coming energy crisis, or if it had you already standing in line to get your tickets to the next No Nukes concert, the nuclear bout is on again, and we're just at the very start of the second round.

THE RENEWABLE AGE

Clearly we have not yet either discovered or efficiently implemented our best opportunities for reducing emissions from the sources that provide our world with energy today. Certainly no option presented in this chapter—from clean coal to a resurgence in nuclear power—is going to be a satisfactory stopgap measure to every person as we transition into the new energy age, nor is it going to be popular in the public eye.

But warm homes are popular. Well-lighted streets are popular. Reliable electricity that powers our domestic economies is mighty popular.

What is most disturbing, as we stand on the threshold of another global energy evolution, is that we have yet to insist that our leaders make a coherent and unified effort to develop the technologies that will be necessary to avert the coming global energy crisis—a plan that realistically takes into account and balances human health, environmental integrity, the intricacies of geopolitics, and the growing needs of the world's economic engines.

The best plan, we believe, is a two-pronged defense. First, we should pursue those technologies that show promise to mitigate the ill effects of our current power sources, funding them in order to move them out of the developmental and demonstration phases where they currently reside and into the coal plants and oil refineries where they can potentially do some good.

At the same time, we need to look to the future, to the renewable energy technologies that will necessarily become more and more a part of our everyday lives as oil fields go dry, as concern for the planet's health grows side by side with an expanding global middle class, and as international relationships prove more or less stable. Funding and implementing renewable energy technologies and the infrastructures that support them is the next great and *conscious* step in our energy evolution. Taking the step will require a vision as passionate as FDR's when he formed the REA and a collaboration among governments, industry, and private partners that is broader than any that have come before.

3

RENEWABLE ENERGY, A PRIMER

HAVE YOURSELF A RENEWABLE LITTLE CHRISTMAS

THERE'S A CHRISTMAS TREE FARM IN CHESHAM, A TOWN LOCATED IN THE STEEP green Chiltern Hills of England, that can be a delightful place to visit even if you're not in the market for a holiday tree. The farm is crackling with the energy of youngsters unleashed out of doors, among the evergreens, taking part in the centuries-old tradition of choosing the tree under which they will find their treasures on Christmas morning. Families arrive by the carload to behold a field of snow-filled branches. It's a sight that has inspired the writers of seasonal songs for generations. Cups of hot cocoa in mittened hands, twinkling lights and shining ornaments, not to mention the old, stand-by vision of sugar plums dancing in their heads.

The Chesham farm is a member of the British Christmas Tree Growers Association (BCTGA), an organization that has taken pains to assure that the visions conjured by its members' products aren't displaced by a press that, from year to year, promotes artificial Christmas trees as the more environmentally sound holiday alternative or environmental groups that advocate having no tree at all as the only eco-responsible option.

While forgoing the family tradition of a Christmas tree is something that even treehugger.com concedes is not going to be generally embraced anytime soon, the BCTGA has commissioned its own study to counter the argument that artificial trees are a more viable eco-option than the real thing. It found, of the brands of trees it analyzed, that all of the plastic parts of the artificial tree were made from polyvinyl chloride (PVC), an especially durable plastic that scientists predict can never disintegrate. But they also note that, however durable its parts, the average life span of an artificial tree is five years—five years in your home, the BCTGA tells you, and uncountable centuries in a landfill! Moreover, the trees used for the study were all manufactured in China. The organization rightly included the fossil fuels used to ship the trees in its environmental analysis of artificial evergreens.

The purchase of a real tree, the association maintains, actually contributes to environmental preservation. Christmas tree farms, they say, are a valuable method of the terrestrial carbon sequestration we wrote about in chapter 2. The tree farms act as carbon sinks, with one acre of trees soaking up the carbon emissions for the equivalent of eighteen people daily. In the United States, where half a million acres are given over to Christmas tree farms, that would be like neutralizing the carbon output of 9 million people each and every day. And because Christmas trees are harvested every seven years or so, and each mature tree is replaced immediately with a new, carbon-hungry sapling, Christmas tree growers would indeed seem to be part of an industry whose economy inherently depends on the best practices of environmental standards.

Not so fast. Remember that Christmas tree farming is a business. Tree farmers can expect an annual gross income of between $600 and $1,000 an acre; they have an industry that in one small country alone, Denmark, does $204 million worth of business worldwide annually. Naturally tree farmers are going to put the best spin on the business. Tree farms are carbon sinks, of course, but that environmental benefit may well be negated by other agricultural practices currently widespread among tree farmers.

The species of trees, for instance, that are most desired during the holidays are also some of the most susceptible to infestation, so farmers use pesticides to protect them, and the pesticides end up contaminating the local water supply. Frequently helicopters burning fossil fuel are used to lift the hewn Christmas trees out of wooded acreage and drop them at the point of sale. If Christmas trees aren't disposed of properly—converted to compost or mulch, or sunk into a pond, or acquired as rooted trees that can be replanted outdoors as soon as the holidays are over—they themselves become environmental liabilities.

Finally, fir trees are not among the most carbon-efficient species. Would the land taken up for Christmas tree growing better serve the environment if it were planted with some faster-growing, carbon-hungrier species, like poplars? Is the claim of Christmas tree growers that their acres of firs are a benefit to the environment just wishful thinking—or good marketing—by another polluting industry?

DEFINING OUR TERMS

"When I was a girl," says a friend who grew up on a farm in the American Northeast in the 1940s, "we got our Christmas tree from our own backyard. The day after Christmas we'd undecorate it and drag it outside. The winter birds had a feast on the strings of popcorn and cranberries we left on the tree for them and, sometime later, my father would chop the tree up and we'd use it to feed the fire in the living room fireplace. On a farm, you didn't waste anything. The Christmas tree was just part of the mind-set of thrift—you used every part of it."

Grow it, heat your home with it, make furniture out of it, decorate it for the holidays, use every part of it, grow more of it—wood is *the* classic renewable energy resource. But it's not *clean* energy. Burning wood for fuel releases the carbon stored in the trees into the atmosphere. For most of the time that humans have used wood as fuel, new tree growth neutralized the output of carbon from burning wood by sucking the released carbon back out of the air. Wood is still renewable, of course, and many new renewable energy technologies use wood or wood products, or the stored energy of other agricultural products, to make fuel and power. And all these often are good choices in this context because wood is the lowest carbon emitting fuel of all.

But because renewable energy is frequently defined as what it's *not*—that is, it's *not* a fossil fuel—it's important to get straight on what it *is* before we really start to talk in depth about it. Words like "renewable," "clean," and "sustainable" are often used interchangeably, but they each have their own subtle distinctions.

Renewable energy comes from those resources that can be replenished in a relatively short period of time. They include biomass—wood, switchgrass, and other agricultural crops—hydropower, geothermal energy, wind energy, and solar energy. *Clean* energy comes from those sources of power and fuel that don't produce carbon as they produce energy, and that excludes wood.

Sustainability refers to a process or state that can be maintained at a certain level for an indefinite amount of time. For planet Earth, this means using our resources and serving our environment in a way that benefits us now as well as into the indefinite future. One of the most often-cited definitions of sustainability is the one created by the Brundtland Commission, led by former Norwegian prime minister Gro Harlem Brundtland. The commission defined sustainable development as development that "meets the needs of the present without compromising the ability of future generations to meet their own needs." Wisely managing our forests and other natural lands—and even our tree farms—is an example of the process of sustainability.

That process or state of sustainability is accomplished through *conservation*. Energy, as we noted in chapter 2, is "the capacity for work"— the amount of work it takes to perform a task. According to the law of conservation of energy, energy cannot be created or destroyed. When energy is "used," it is merely transferred into a different type of energy—as your car engine, for example, burns gasoline, it converts chemical energy into mechanical energy. The total amount of energy in the universe doesn't change.

What conservation calls on us to do is use the source of energy that will do the most amount of work with the least amount of resources spent. Taking the bus to work spends less of the finite resource of fossil fuel than driving a car; riding a bike, when that's possible, spends less energy than taking a bus. Taking one suitcase when we travel on a plane lightens the load so less jet fuel is spent than if we'd taken two suitcases. Pushing a rotary lawnmower rather than revving up a power mower uses no fossil fuel energy at all. Human energy—*muscle* energy—is remarkably renewable, through eating and sleeping, and it is the ultimate clean energy. It's also a healthy way to spend a resource: We have a friend who dismissed her lawn service and started pushing a hand mower around her own sizable lawn. "I'm saving money, gas, pollution, and"—she winked—"I've lost twelve pounds and two dress sizes without doing a thing other than walking around my lawn once a week!"

As our friend pushes her hand mower, or as an amount of gasoline or coal is burned, or as a trade wind turns a rotor on a turbine activating a generator that sends electrical current to Ben's Chili Bowl on U Street so the restaurant can heat up your order of the house specialty, potential energy stored in muscles, ancient vegetation, or the passing breeze is liberated. This liberated energy is measured by scientists in minute increments—a *joule*, which is equal to 1 *ampere* passed through a resistance of 1 *ohm* for 1 second. Sometimes we refer to an amount of energy as a quantity of Btus, a Btu being a measure of the heat content of fuel. But most of us are more familiar with measuring electrical output in terms of watts.

A *watt* is a unit of power equal to 1 joule per second—or what we know as the capacity of the 25-watt light bulb that's in your reading lamp right now, illuminating this page at the rate of 25 joules per second, slightly less brilliant than a 60-watt bulb but conserving 35 watts of energy that, perhaps, your daughter is using this very same second to go online and check her evening's homework assignment on her school's Web site.

A *kilowatt* is a unit of power equal to 1,000 watts. This is how power use is measured in homes and small businesses. If you live in the United States, you can expect your family's monthly residential bill to reflect on average a charge for 938 kilowatt-hours of power. If it's December, and you've put up a Christmas tree and decorations, you can expect your lights display to draw an additional 136.4 kilowatts over the month of December and, accordingly, increase your electric bill by around $13.

Here's a reference to keep handy as you continue to read: Your additional expense of $13 for holiday lights stays pretty much the same no matter what sort of fuel your electric provider uses at the power plant. That's because government subsidies, as well as government guidelines regulating the use of each sort of fossil fuel, help to balance the expense of a cheap fuel like coal with a costlier one. It costs between 2 and 5 cents to generate 1 kilowatt-hour of electricity when using coal as a fuel. One kilowatt-hour costs about 6.5 cents when it is produced by natural gas, and about 9.5 cents when it's produced with oil. One kilowatt-hour of electricity, when it comes to your house from a nuclear reactor, costs around 2 cents.

The electric use of larger businesses and institutions is measured in *megawatts*, a unit of power equal to 1 million watts. Pennsylvania State University, for example, sits on a main campus of 15,000 acres and educates 42,294 students a day, and needs approximately 46.77 megawatts of electricity a year for all of its classrooms, science labs, and athletic fields. Some of that electricity, as a member of the Environmental Protection Agency's Green Power Partnership, comes to the university from renewable energy sources. In China, the factories that produce plastic merchandise—toothbrushes, eyeglass frames, artificial Christmas trees—use annually a part of the 3.3 million megawatts that drive the nation's industrial sector.

Taking it up a notch, the energy consumption of states and nations, and the output over time of power-generating plants, is measured in *gigawatts*, units of power equal to 1 billion watts. For comparison, the United States, with a population of 296.68 million, required 4,046,600 gigawatts of power to sustain its business and pleasure in 2006; China, with a population of 1.3045 billion, required 2,322,720 gigawatts; England, with 60.22 million people, 376,630 gigawatts; Brazil, with 186.41 million people, 375,190 gigawatts. The portions of power each country drew from renewable energy sources in 2006 are, respectively, 9.11, 16.83, 5.35, and 88.9 percent. Worldwide, the total amount of energy generated by renewables is about 18 percent. Perhaps startlingly, that percentage is expected to decrease to 16 percent by the year 2030 because of the growing global demand for electricity. But there is a vision, presented by the American Council on Renewable Energy (ACORE) that is much more heartening: The organization estimates that, with helpful government policies in place, renewable energy technologies can supply 635 gigawatts of generating power in America alone by 2025. That's more than double the new capacity experts forecast will be needed by then.

There is one other way in which watts are measured, and that's terawatts, units of power equal to 1 trillion watts. Why do we need terawatts? If the energy consumption of *nations* is measured in gigawatts, what need do we have for a term that measures a larger entity? What larger entity could there be?

In the United States, holiday lights draw, on average, 2.22 terawatts of electricity annually.

It's important to have some idea of the enormous numbers we're talking about as we move into our discussion of how renewable energy sources and the infrastructures that support them can be integrated into the world's energy mix. It will be helpful in seeing how renewable energy can impact those numbers. And, we think, it helps to make these enormous numbers easier to conceptualize when we can fit them into the context of the small percentage each one of us might be personally responsible for using, or conserving.

But there's one more thing that isn't usually calculated into the equation when electricity is counted out in watts and dollars, though we think it ought to be: the savings that using renewable energy can bring to the costs of human health care. Remember that in the process of generating a billion Btus of energy from coal, 208,000 pounds of carbon dioxide are released into the atmosphere. The process of generating a billion Btus of energy from renewable sources—with, for instance, wind power—releases absolutely *zero* greenhouse gases into the air. As a matter of fact, generating 2 billion or even 1 trillion Btus of energy with wind power produces zero greenhouse gases.

Human health, as we all know, is directly affected by man-made pollutants. Asthma, immune disorders, and a panoply of cancers, to cite just a few diseases, have been linked to pollutants in the environment. But the cost of health care for the people who are exposed to the pollutants has never really been a part of the cost-benefit ratio used in assessing or approving potential power projects. Today that's starting to change. Experts who analyzed the impact of the AEP settlement, for example, estimated the cost of health care could be reduced by as much as $32 billion.

That's $32 billion saved in insurance company payouts, Medicare and Medicaid payments, co-payments, and other personal medical bills.

In Europe, according to the European Environment Agency's 400-page report issued in October 2007, pollution is to blame for hundreds of thousands of people across the continent dying prematurely. Air pollution, the agency reports, has cut Europeans' life expectancy by nearly a year.

In the same month, however, that the European Environment Agency released its report, something historic happened in the United States. The Kansas Department of Health and Environment denied permission for the construction of a new coal-fired electricity-generating plant. This marked the first time a government agency rejected a proposed power plant for public health reasons. The broad vision it will take to help us transition into the renewable age will necessarily calculate the cost of health care into the cost-benefit analysis of power—and, in Kansas, we see the vision is already as broad as the state's vast, golden fields.

GREENWASHING

There's one more term that we ought to define while we're at it, and that's *greenwashing*. Greenwashing is what happens when a hopeful public eager to behave responsibly about the environment is presented with "evidence" that makes an industry or a politician seem friendly to the environment when, in fact, the industry or the politician is not as wholly amicable as it or he might be. We touched on this concept when we talked about the Christmas tree-growing industry presenting partial evidence of its ecobenefits—tree farms as carbon sinks—while neglecting to mention the polluting pesticides or harvesting helicopters. Greenwashing is a marketing strategy, and one the public might grow ever more susceptible to as our need for energy expands and the CO_2 in our atmosphere continues to accumulate. As we grow ever more anxious for answers to our energy problems, we need to foster a healthy skepticism and understand that some of the answers that result won't be wholly reliable.

Let's use the commercials for hydrogen-powered cars that are starting to make appearances on television as an example. BMW, Honda, Ford, and Mazda are some of the carmakers whose research and development of hydrogen-fueled cars was begun early on. Therefore, these companies are likely to be the leading manufacturers in bringing this new technology to the market.

So far, so good. Few people dispute the potential benefits of hydrogen technology. Hydrogen-powered cars produce no tailpipe emissions of greenhouse gases or other pollutants, which augurs well for climate change concerns, and using hydrogen as a replacement for oil speaks eloquently to the need and desire for energy security. What's more, engineers are optimistic that these cars of the future can offer the consumer roomy interiors, sleek design, high efficiency, and good performance. Even better, right? When can we take delivery? Commercials even now seem to prepare viewers for the imminent introduction of these hydrogen-powered cars; the commercials are, however, at best misleading.

First of all, let's talk about the process of making hydrogen fuel. There is no "alchemy" involved in hydrogen technology, as one carmaker claims. It is a function of hard science that isolates hydrogen molecules and allows them to be transferred into the energy needed to run a car. Although the idea of car emissions amounting to no more than pure water and steam may seem like magic—and, indeed, that's all emissions from hydrogen-powered cars amount to—dirty work happens in isolating the hydrogen molecules in the first place. Making hydrogen fuel takes energy, and right now that energy comes from—you guessed it—coal or oil or natural gas.

This doesn't mean that in time the power needed to isolate hydrogen molecules can't come from wind energy, or solar energy, or energy from biomass. It does mean that hydrogen technology isn't yet mature enough to allow energy that originates from these sources to make the production of hydrogen fuel truly renewable, environmentally friendly, or cost efficient.

How much time do we need to do that? A commercial I saw last night for a hydrogen car promised that the vehicle is ready for the world, as soon as the world is ready for it. But according to Joan Ogden, a professor at the Institute of Transportation at the University of California, Davis, it will be "several decades" until the world is ready.

Why such a long time? A little Googling and it's easy to find out that hydrogen cars aren't exactly ready for the world either. Honda, for instance, isn't planning on selling its hydrogen car any time soon. It will be three or four more years until a version of its hydrogen fuel cell vehicle, the Honda FCX, will be rolled out, and then it will be available only in Japan. The car will make its debut there because that country already has a dozen or so hydrogen fueling stations. Along with offering the consumer reliable performance from a reasonably priced hydrogen-powered car, you see, a hydrogen refueling infrastructure has got to be built so that the fellow in Cleveland who buys a Ford hydrogen car, and the gal in Madrid who buys a BMW hydrogen car, and the guy in Tokyo who takes delivery of his Honda FCX each has got a place to go to fill up the tank conveniently and at a competitive price.

Honda is taking an original tack to address the lack of hydrogen fueling stations around the world. It's developing what it calls a "Home Energy Station." Hydrogen car buyers would, in theory, purchase one of these stations along with their car and use it to make hydrogen fuel right in their own garage.

A fuel station right in my own garage? It seems too good to be true! Well, it is.

The "Home Energy Station" is designed to be powered with—here it comes!—natural gas!

The problem of how to create clean hydrogen fuel without the use of dirty fossil fuel is just one of the ongoing technical issues carmakers face. While hydrogen cars currently are part of experimental fleets for some government agencies—the state of New York has been using them on a very limited basis since 2004—other obstacles that stand in the way of commercial versions of the vehicles include where to put the cumbersome onboard tanks needed to store the hydrogen fuel and how to assure reliable cold weather startup in a vehicle that produces water as one of its by-products—water that turns to ice when the temperature dips below 32° F!

As Professor Ogden says, hydrogen-powered vehicles could be an important part of the "larger trend toward decarbonization of energy and more efficient use of resources." But that will take time. And it will take money.

Real money.

In January 2003, President Bush announced his Hydrogen Fuel Initiative. The president, according to a White House press release, "envisions the transformation of the nation's transportation fleet from a near-total reliance on petroleum to steadily increasing use of clean-burning hydrogen." But as we've just seen, there are barriers that prevent this vision from becoming a reality in the near future. Let's review.

The first barrier we've already talked about. Hydrogen won't be a "clean-burning" fuel until its technology allows us to stop using a fossil fuel to produce hydrogen fuel. As it stands now, that's just transferring using fossil fuel in cars to using it at some earlier point in the transportation chain.

The second thing we have to question is the amount of money being committed to the development of a "hydrogen economy." Depending on how you break down the numbers cited, the U.S. commitment to the Hydrogen Fuel Initiative is anywhere from $2.9 to $3.6 billion over a period of five years. Now, any one of us, our whole large extended family, our five dearest friends, our closest neighbor, and the horse he rode in on could live exceptionally well for five years on $2.9 billion. But let's put $2.9 billion into perspective as a reasonable amount of funding for a project that is supposed to build the foundation for a whole new economy—reverse climate change, avert an energy crisis, and enhance national security by removing our dependence on foreign oil: The United States currently spends $195 million *a day* fighting a war in Iraq, blowing through several billion every few months. In a little less than two months, more funding is poured into this military action than is devoted to a solution to our energy crisis in over five years. In contrast, California, one individual state, has pledged to spend $3 billion to develop its solar power capacity, just one type of renewable energy. If we intended to be serious about the dream of a hydrogen economy, we'd have to equip researchers with serious money so they could work more efficiently, and more expeditiously, to develop the technology to its maturity.

Now let's consider one final problem with the hydrogen economy, the third obstacle we'd have to surmount to make a hydrogen-fueled world possible: The president's Hydrogen Fuel Initiative doesn't call for the consumer to have a choice about juicing up a hydrogen car until 2020. That timeline fits in well with the estimates of researchers and carmakers for bringing hydrogen technology to maturity. But we can't wait until 2020 to start making the changes that will reverse climate change, avert an energy crisis, and enhance national security.

Hydrogen technology, for all of its real potential benefits, is not a near-term solution to our energy challenge. We need to be able to make realistic choices now while planning for our long-term energy future.

Our first choice has got to be to pass legislation that increases fuel efficiency standards for cars. Plain and simple. Without question, that is the first step that will provide us with immediate and important relief.

Next, we need to form meaningful collaborations among governments, industry, and private partners to develop and implement the renewable technologies that will provide us with clean energy as our sources of fossil fuels dwindle. This means a sustained conviction to supply consumers with electricity generated with the power of the sun, home heating generated with geothermal power from the earth, fuel for our cars made from plants that grow, as well as the ability to manufacture hydrogen fuel using the power of the wind.

Focusing myopically on hydrogen technology as the sole answer to our energy challenge is wasteful of time, of the finite resources we do have in hand, and of the mature renewable technologies that already exist. Meeting our future energy needs will require a variety of strategies. Fortunately, we have a real abundance of viable alternatives. In the next three chapters we talk in depth about why it's necessary to pursue diverse strategies, and just what all of those strategies are.

4

POWER PLAYING

IN 1989 A COMPANY CALLED ELECTRONIC ARTS INC. (EA) INTRODUCED A computer game called *SimCity*. To play the game, one creates "Sims"— *sim*ulated people—choosing their costumes and facial features as well as their personalities (sloppy or neat; shy or outgoing), their personal goals (some Sims want to be really, really rich, others only long to fall in love, and still others want to be the smartest Sim on the block), and one constructs the homes and businesses the Sims live in and among, in styles that can range from medieval castle, to modernist dream, to barely habitable hovel. The object of all of this is to assist each Sim in achieving its own particular aspiration by furnishing it with a pleasing personality, appropriate education, and a strong support system of other Sims—or to watch as its life falls apart because of bad choices.

Neither of these outcomes is as easy as you might think. It seems, to the novice, that the Sims game player is omnipotent. But even the Sim with the most advantages may find herself one hasty decision away from unemployment, and the Sim struggling in the most inhospitable circumstances is pretty darned tenacious about finding ways to survive no matter how many times it tries to cook a meal that's beyond its skill level and, as a result, sets its kitchen on fire.

Since its launch, SimCity Societies has become one of the world's most popular electronic game franchises, with versions that set the Sims in the big city or on a college campus, or allow them to buy pets or to mix it up with paranormal characters. More than 18 million copies of the games have been sold throughout the world and, in 2007, EA posted revenues of $3.09 billion.

What has all of this got to do with renewable energy?

BP, formerly known as British Petroleum, is one of the world's largest energy companies. Over 96,000 employees look after its business in over 100 countries. BP was also one of the first traditional energy companies to recognize the coming global energy crisis and to take significant actions to help to avert it. In 2005 it launched BP Alternative Energy with an $8 billion commitment to building power plants that generate clean energy from such sources as wind and solar. The company also has a lot to do with one small computer game that retails for $49.95.

Knowledge is power. Educated young people have historically been an integral part of social progress. When BP wanted to help educate young people about the changes in energy choices that we are going to have to start to make, it knew that the best way to deliver the message was through entertainment, so it got in touch with the folks at EA, a company that has $3.09 billion of street cred with the younger generation. The result is the newest iteration of SimCity Societies, released in November 2007, which allows game players to make choices about how their Sims community is going to be powered.

Just as in real life, one power choice in the SimCity universe is to use coal to create electricity, and, just as in real life, in SimCity Societies coal is plentiful and the electricity in Sim houses and businesses is cheap. Game players can conserve their Simoleons, the in-game money, by building coal-fired power plants to run their SimCity. But if they do, real-life consequences will kick in: Carbon levels rise, environmental protestors hit the streets, and the communities must deal with climate change disasters like heat waves and droughts.

Game players also can choose to build green, powering their cities with solar fields and wind farms. In this case, the property values in the SimCity remain high, the Sim citizens are safe from natural disasters, and the Sim characters, because they are exposed to less pollution, get sick less often and so miss less days of work.

But just as it is in the real world, the renewable energy plants also cost game players more of their Simoleons, don't produce as much power per dollar invested as higher-carbon-emitting alternatives, and take up more of the Sim community land. Thanks to BP's input, SimCity energy cost analyses are real-world accurate, and carbon emissions reflect real-world ratios of carbon dioxide (CO_2) and the corresponding consequences. Factual snippets about energy conservation appear within the game to assist players in managing their fictional global energy challenge. Players can even opt to build a Carbon Exchange in which they pledge to cap their carbon emissions and are rewarded for coming in under their limit, simulating the sort of carbon offset and renewable energy certificate programs we talk more about in chapter 7.

We like this concept as a way to get kids knowledgeable about and involved in their energy future a great deal—conveying facts from the experts at BP ingenuously, through the fun created by EA. We also think this kind of game, which challenges its players to think critically about energy options and explore creative solutions to our energy challenge, isn't just for kids. Grown-ups too could benefit from the opportunity to stretch their energy imaginations.

Certainly, being creative about energy conservation and alternative energy options isn't a game just for politicians and CEOs either—as we've already pointed out, the energy revolution is a bottom-up movement: It's all about people like us pushing our leaders in the right, creative direction. That's why, in our version of the game, the reader becomes the leader—and at a very local level. We'd like to invite you, for this and the next two chapters, to sit in as the mayor of your town.

In the real world, national governments are struggling to meet their energy goals. In the European Union, for example, the goal is to raise the share of energy created from renewable resources by 20 percent, and to reduce carbon emissions by 20 percent, by the year 2020. But it is small communities that are proving to be the most successful at sustainable living, frequently far exceeding the targets of their national leaders.

On El Hierro, the smallest of the Canary Islands, a 107-square-mile speck in the ocean, a revolutionary wind/hydro plant is being developed that will allow residents to rely on renewable energy alone within the next few years. The plant is being engineered to continue producing energy even as weather conditions fluctuate. While the powerful trade winds blow, extra wind energy is used to pump water into a reservoir that lies 2,300 feet above sea level; when the winds are calm, the stored water is released to a lower reservoir to produce a steady flow of electricity.

On Samso, an island of the eastern coast of Denmark, the 4,200 residents are even further along in the renewable race. Every watt of the island's electricity already comes from just eleven wind turbines, and 75 percent of its heat is provided by renewable sources, such as solar and biomass. In southern Sweden, the city of Vaxjo, located amid lush forests, is converting to biomass-based heating by burning wood waste in its power plants—wood chips and pellets; this city of 78,000 can already boast that 52 percent of its energy supply comes from renewables and that it has reduced carbon emissions by 30 percent *from 1993 levels.*

"When you start with a top-down approach and tell every country they have to reduce carbon emissions by 20 percent, they don't know what to do," according to Pedro Ballesteros Torres, head of the European Community's Sustainable Energy Europe campaign. "But if you start with a very democratic and bottom-up approach, it's not so ambitious and not so impossible."

In the United States, where local and regional leaders are frustrated by the lack of action on energy issues in the nation's capital, the race toward a renewable future is being run aggressively at state level. California, New Mexico, New Hampshire, Vermont, Massachusetts, Connecticut, and eleven other states so far have established their own carbon-reduction targets. In the Midwest, nine governors, along with the premier of Manitoba, signed an accord in November 2007 to establish the first state caps on greenhouse gas sources from factories and power plants as well as a system to trade pollution credits that could reduce emissions by a minimum of 60 percent.

In fact, asking you to sit in as mayor has its roots in a very heartening reality. In 2007 former president Bill Clinton announced that the William J. Clinton Foundation is partnering with the U.S. Conference of Mayors to help 1,100 cities gain access to volume discounts on energy-efficient and clean-energy products offered through the purchasing consortium the foundation has created with retail giant Wal-Mart.

This is just one way that mayors are making a huge difference in how the people of the United States use and think about energy. In 2005, at the U.S. Conference of Mayors' annual meeting, 141 mayors

signed on to strive to meet or beat in their own towns and cities the targets of the Kyoto Protocol that the U.S. federal government had
declined to ratify. In May 2007 Mayor Kathy Taylor of Tulsa,
Oklahoma, became the 500th mayor to sign the agreement.

It's because of leadership like this, in states, cities, and small towns
currently powering the energy evolution—surpassing, in many cases,
the commitments and creativity of those in positions in national
governments—that we'd like to invite you to step into a municipal leadership role. Whether you are a private citizen or, in fact, a world leader,
taking a mayor's-eye view of the energy status quo around you, and
imagining how it can be improved upon on a small scale, helps to bring
solutions into manageable perspective.

Are you game? All right, then. How do you play?

Think of this chapter and the two that follow it as your playbook—an
overview of the technologies that you can employ to meet your mayoral
goal of transitioning your municipality into the new energy age. To
streamline an understanding of your new energy options, we've separated
the discussion into three major categories that you'll need to address:

1. Renewable *power*—that is, the sources of energy that can provide your city or town with clean electricity: solar power, wind
   power, hydroelectric power, geothermal power, and power
   from biomass—which will be the subject of this first chapter.
2. The power *grid*, or how you can most efficiently deliver clean
   electricity to the homes and businesses in your community.
3. Renewable *fuels*, ethanol and biodiesel, and information
   about how they can best be delivered to your constituents at
   the filling station.

The object of the game is to transition your community from traditional
sources of power and fuel into 100 percent green energy use. These
chapters will guide you as you figure out the energy needs of your
particular community and the best local resources to tap to meet it.

CALCULATING YOUR ENERGY CHALLENGE

Let's talk about size. How many people live in your town? Remember that the average household in the developed world draws about 950 kilowatt-hours of electricity a month from the grid. In a town of 40,000 people, assuming an average of four persons per household, you might then need to secure 9,500,000 kilowatt-hours a month for domestic use alone. How are you going to generate that electricity?

What businesses in your town also need electricity? How much power does your favorite diner use every day? How much power does it take to produce your town's daily newspaper? How much electricity does it take to run the salon where you get your hair done, the office of the CPA who does your taxes, the shopping plaza where you buy your groceries and get your prescriptions filled?

Where do the people in your town work? What is your area's largest employer, and how much electricity does that industry require to keep its supply chains moving smoothly? How many municipal buildings, town halls, or courthouses does your community support? How many school buildings?

What about fuel? How many cars do the people in your town own? What kind of cars are most prevalent? What is the average daily commute to work?

Being mayor is a complicated business, not all cutting ribbons and kissing babies. You can, of course, do a lot of research—telephone calls and Internet searches—to find out specific answers to the questions we've posed, and if you were really an elected official, that would be exactly what you'd do. For the purposes of the game, you can keep it simple and work with these figures:

The average individual living in a developed country uses about 12.5 kilowatt-hours of electricity every year in the process of home and business life and about 464 gallons of fuel annually.

Just multiply the average use by the number of people in your community to figure out the scope of the challenge you'll be dealing with as mayor.

Next, as mayor, you'll have to consider both the natural geographical features and the existing man-made assets of your town. Is it built on a waterway that would make hydroelectric power a good energy choice? Is it surrounded by open plains, or high mountains, that would make wind power especially viable? As you read through the playbook, let your mind roam over the ways in which each technology might be adapted to the particular features and assets your city already enjoys. You'll likely be pleasantly surprised at how many options are actually within your reach, and at how several sources of renewable power, deployed in complementary ways, can be of use in fulfilling your community's energy needs, as the way the people of El Hierro have combined wind and water power to become energy independent.

Finally, but not least, among the things you'll have to do as mayor if you're going to win the energy challenge is to leap a big hurdle we talked about earlier: reorienting yourself to how energy will be produced in the near future and educating your constituents so they can do the same. Here's an interesting take on how to keep yourself balanced in the process of the big jump.

Our grandmothers cooked with lard, wore corsets and girdles with complicated stays and laces, and smoked Camels because they were "recommended by more doctors than any other cigarette." Another generation of women teased their hair into bouffant helmets, believed that red meat was essential to proper nutrition, and were advised by physicians who prescribed amphetamines to control weight gain. The women who managed yesterday's households stoked coal furnaces or thought of long and meandering Sunday drives as cheap and wholesome family entertainment; for many of them, the entire concept of recycling was embodied by the old coffee tin where the bacon fat was stored.

Times change.

Fashions change. More precisely, scientific research and social progress abet each other to improve our lots. As we have turned away from old-fashioned staples like lard that clogs our arteries, we're turning away from the fossil fuels that clog our air and toward more healthful alternatives.

But let's not throw out the baby with the bathwater. Fashions don't change out of the blue; even the most cutting-edge designer on Seventh Avenue looks to what has come before for inspiration—or how else in heaven to explain the return of corsets?

Our grandmothers hung out their laundry to dry in the breeze—a practice many of us have again taken up as an electricity-saving measure and that has, at the same time, returned to us the singular pleasure of sleeping in crisp, sun-scented sheets. They conditioned air by throwing open the windows after the sun went down and then closing them again and drawing the curtains first thing in the morning to preserve the night's coolness indoors all day long—an easy sweep through the house that can reduce or even replace the need to turn on our air conditioners today. They caught rain in barrels and used the water for the vegetable garden and flower beds instead of turning on an outdoor spigot. They took advantage, in short, of renewable resources, as people the world over have been doing since ancient times.

Renewable energy is nothing that's brand new. Advances in technologies simply allow us to tap into these resources on the much grander scale that the blossoming economies of our burgeoning population require. By putting ourselves into the position of someone like a mayor who has got to take responsibility for how new energy will be supplied to a very specific population and by imagining energy solutions at a very local level, we can all become more deeply engaged and empowered to accomplish the new energy reality.

Solar Power

Ra, Shamash, Helios, Apollo: Sun gods and sun worship were major parts of religious ceremonies in cultures from ancient Egypt to England, Mesopotamia to Mexico, China to Native America. The sun has been associated with benevolence, prosperity, fertility—both human and agricultural—and, in acknowledgment of the ancient world's understanding of the sun's centrality to their civilizations, it has even been equated with justice.

Today we know the sun as simply our nearest star, and that without it life on our planet would cease to exist. As we noted previously, the solar system is still our most fundamental source of energy. Fossil fuels are, at the end of the day, nothing more than the stored solar energy of prehistoric vegetation; even nuclear energy, that most modern of all energy sources, has its roots in the heavens: The uranium that is used to fire nuclear reactors was created by novas—exploding stars.

Technology has surely expanded the ways in which we can access the sun's energy. In the Mojave Desert, for instance, at Kramer Junction in California, 1.2 million mirrors glitter in parallel rows over 1,500 acres, reflecting and concentrating the sun's power in the largest solar field in the world. These mirrors provide 350,000 homes with electricity—and evidence of the continuing centrality of this brilliant star to our lives.

As mayor of your town, there are two ways of harvesting the sun's power that you'll want to consider in devising your energy plan: *passive* and *active*. In the *passive* sort, no mechanical equipment is used to access the sun's power. Like many of the things we do nowadays that we think of as so innovative, this method of solar collection has a long-established history. In the days before electrification, passive solar systems were a common way of heating water—much more convenient and economical than hauling buckets of the stuff to boil over a wood or coal fire.

Modern passive solar heat systems are often used today to heat water for swimming pools. The systems use panels made with silicon, a space-age material made from ordinary stone-age sand, to focus the sun's heat on pipes that are filled with water. When the condensed power of the sun has worked to heat the water in the pipes, that water is funneled into the pool.

Perhaps, as mayor, you'll propose a regulation that requires all pool owners in your town to install such a passive system—and, then, because you want to be elected to a second term, you'll do two other things. You'll give a speech that outlines the quantity of fossil fuel-generated electricity and corresponding CO_2 emissions the passive pool heating systems will save, and you'll enact an incentive program that helps pool owners to afford the systems through direct subsidies or maybe tax breaks.

Active solar power involves the use of fans and pumps and other mechanical devices to direct the sun's power in space-heating applications and to generate electricity. There are a large number of competing active solar technologies—we won't go into all of them; rather, we'll talk about the mature technologies that are already in widespread use as well as some emerging developments that hold promise of being the next most viable advances in the field.

Photovoltaic Technology

Photovoltaic (PV) technology is what we're referring to most often when we talk about the rooftop solar panel systems that have become more and more common over the last fifteen years. Many of us are actually more familiar with this technology than we realize—thin, flat-plate, light-sensitive silicon cells are what trip on those little calculators so many people whip out when they need to figure out the tip after dinner at a restaurant. It's this same technology that is finding its way to our rooftops.

Flat silicon panels are placed in a fixed position on the roof, usually facing south and sloped at an angle to the horizon equal to the latitude plus 15 degrees so that the panels are in an optimum position to collect the most sun power. The flat plates intercept and absorb the sun's rays. These rays then pass through a transparent cover, on their way to a heat-transport material, such as water or air, that flows through tubes on the underside of the panels. The water or air removes the sun's heat from the absorber and is then circulated with pumps or fans to provide warmth in space-heating applications.

PV panel systems can also be used to make electricity. In this case, the fans and pumps direct the heat that's extracted from the sun to a generator. The resulting electric power can be used to run the lights and appliances in the individual home or business upon whose roof the panels sit—and it can also be connected to the grid from which the individual structure draws more traditional power.

It's this connection to the grid that can make these private solar-generating systems extra efficient and cost-effective. By connecting the solar-generating power from your house to the community grid, you actually can become an energy provider, feeding the electricity you don't use in the course of the day's sunshine back into the supply that's available to your energy company to direct to other customers. You get credit for supplying this electricity, and then the most amazing thing happens: Your electric meter actually works *backward*.

But here's the downside: It can be expensive to install solar panel systems, and the more sophisticated the system—passive to active and grid-connected—the greater the expense. Though the home or business owner will recover the cost of the solar installation through savings on monthly electric bills, and the system will eventually pay for itself because of these savings, this recovery occurs incrementally, over a period of years, and it's often a stretch for an individual homeowner to work the average cost of $16,000–$20,000 for a grid-connected solar installation into the annual household budget. Stepping up to assist families and small businesses in acquiring solar panel systems is one of the ways governments around the world can help—and are helping—to assure their citizens a supply of electricity that's reliable and adequate to meet their needs as well as wholesome for the environment.

As mayor, some of the solar installation programs already in place may inspire you with ideas about how to go about turning the rooftops in your town into solar systems. In Australia, for example, a $75.3 million "Solar Cities" initiative is now under way with the goal of accomplishing the mass installation of PV power. In Spain, BP Solar has partnered with one of the country's leading banks, Banco Santander, to jointly carry out the largest investment, public or private, of PV power to date in Europe. In China, solar panels are being used to bring 100 megawatts of electric power to remote rural villages where the installation of more conventional sources of electricity has proved logistically impractical.

In the United States, a program called the Million Solar Roofs Initiative has provided federal grants that resulted in the installation of 200 megawatts of grid-connected PV capacity, 200 megawatts of solar–water heating capacity, and an estimated decrease of 3.3 million tons of CO_2 emissions. By the measure of goals outlined within the Million Solar Roofs program itself, the initiative has been a real success.

But the program provided for just $16 million in funding and ended in 2006. Though the 109th Congress, in one of its final acts, extended a 30 percent solar energy investment tax credit, that too is set to expire on December 31, 2008. If, in your position as mayor, you grow impatient waiting for federal constancy, you could initiate a program on the order of what Berkeley, California mayor Tom Bates and his City Council approved in November 2007. Back in 2006, Berkeley had already passed a measure setting reduction targets for its city's greenhouse gas emissions and directing the mayor to develop a plan that would meet those targets. The plan the city has come up with includes solar power. Berkeley's solar roof program is, in fact, the first of its kind in the nation, and it stands to increase the 400 solar rooftop installations already in place in the city by the thousands. Berkeley will do this through the creation of a "sustainable energy financing district."

The way this financing district works for the business building or homeowner who wants to install a solar panel system is pretty simple. The individual property owner contracts with an approved solar contractor. The city provides the funds for the project from a well-secured bond that offers lower interest rates than are commercially available. The property owner repays the city's investment over twenty years, through an assessment on his annual property tax bill.

This arrangement has many benefits. First, the solar installations the city helps to fund enables it to meet its CO_2 reduction target. Next, the home or business owner realizes an important improvement to the property that increases its value without a large, up-front cash outlay, through a loan that can be paid back at a lower interest rate than he or she would likely enjoy as part of an individual loan package. Even better, because the repayment is structured through annual property tax assessments, if the property owner should sell before the twenty-year repayment term is complete, the new owner would automatically take over responsibility for the balance of the loan.

Need still another reason, as mayor, to take the development of a solar rooftop program into your own hands? All those new installations in your town will create business for solar panel manufacturers and your local solar panel dealer, and good-paying jobs for workers in your community who will have to be hired and trained to do the actual installations—all of which will quite probably have a positive impact on your local unemployment rolls.

Makes you sort of want to pick up the phone and call Mayor Bates and get all the details right now, doesn't it?

There is one more alternative, for individual home and business owners, to install solar rooftop power systems with less out-of-pocket expense, and that is through one of the innovative companies that are springing up to act as middlemen between solar panel manufacturers and their potential customers. The way these companies—such as SunEdison in Beltsville, Maryland, and MMA Renewable Ventures out of San Francisco, California—work is by purchasing the solar panels themselves outright. They then install the panels in the client's building but maintain actual ownership of them, charging a monthly or yearly "rental" fee to the client. The home or business owner reaps the environmental and economic benefits of the rooftop solar system without having to make a large, up-front cash outlay. Additionally, as part of the rental agreement, the company provides ongoing maintenance for the solar power system, as SunEdison did for its clients after the fires in southern California in the fall of 2007, sending its crews to clean and inspect the panel installations.

This is the kind of business that might have been a natural transition for the small oil jobber we wrote about in chapter 2. Perhaps, Mr. Mayor, there's a small, struggling oil distribution business in your town that could benefit from your city's support in branching out into a new kind of energy—say, in the form of an economic development grant or a low-interest loan?

THIN FILM PV

As research and development of renewable technologies continue there will be, of course, remarkable breakthrough discoveries that improve the way we access renewable power. Much in the same way our friend who collects vintage clothing restores beautiful garments that were painstakingly put together on a treadle sewing machine with her slick, twenty-first-century computerized Swiss model machine, the next generation of renewable technologies will adapt what we already know to make it work for us more efficiently, and at less cost.

The next generation of rooftop solar panels is called thin film PV. These panels differ radically from the rigid, silicon-based models we're all growing used to. A company called Nanosolar, based in California, has developed a method of mass producing wafer-thin solar cells and printing them, as one would print a newspaper, onto aluminum foil. These solar "sheets" require a fraction of the expensive semiconductor material used in the now-common polysilicon PV panels, and they are also flexible—two properties that greatly expand their potential applications. In fact, this aluminum film technique is expected to reduce the cost of accessing solar energy so it can compete with coal-generated power—a detail that makes PV technology, for the first time, practical for utility scale grid operations.

The first thin film PV installations are, indeed, planned for large power plants, rather than for homeowners, with solar stations in the range of up to 10 megawatts in size. Further—as if making solar power cost competitive with coal weren't enough in itself!—these thin film power stations can be up and running in 6 to 9 months. Compare that timeline with the 10 years it takes to make a new coal plant operational, or the 15 years a new nuclear plant requires.

Clean electricity that is cheap and, relatively, immediate—you're going to be hearing a lot about thin film PV electricity, and you're going to be using it, in the months and years ahead.

Solar Thermal Electricity

Solar power today accounts for only .04 percent of all electricity generated worldwide. Making electricity from the sun's energy on a large scale—at the utility scale—requires more complex technologies than do solar rooftop systems. These technologies are grouped under the terms *solar thermal electricity*, or the acronym *CSP*, for condensed solar power. There are three primary solar thermal electricity technologies: parabolic troughs, solar dishes, and solar power towers. As mayor, you'll want to know a little bit about all three of them. Part of your job is, of course, monitoring the amount of electricity used by your constituents' homes, businesses, and industries. You won't want the reserve margins at the plants that send power to your city to fall below 15 percent, and when demand does start to narrow that margin, you'll need to be fully informed about all available types of electricity generation so you can make the best choices to keep your city's power supply stable. Perhaps a solar power plant would fit well into your town's scheme.

Parabolic Troughs

Those 1.2 million mirrors that sparkle in the Mojave Desert are actually highly curved, concave reflective instruments called parabolic troughs. They work by tracking the sun and focusing its energy at thirty to forty times its normal intensity on a pipe that runs down the central point above the curve of each mirror. The pipe contains a heat-transport fluid—often water—that heats to approximately 975° F under the focus of the sun and becomes steam. The steam is directed at a turbine, which, in turn, activates a generator to produce electricity. The cooled heat-transport fluid is then recycled through the solar field and reused.

Solar Dishes

Solar dishes also use mirrors to focus the sun's power. The advantage they have over parabolic troughs is that because of their design—a single, enormous concave mirror—they can concentrate the sun's energy to an even greater degree, heating the transfer fluid up to 1830° F. Like a parabolic trough system, the resulting steam is directed from each mirror setup to a central generator to produce electricity. Also like trough systems, the energy can be collected from any number of mirrors at one time, making both technologies suitable for remote installations—that is, your town doesn't necessarily have to be located near the Mojave Desert to benefit from the electricity that's produced there.

Solar Power Towers

The emerging technology of solar power towers likely represents the next generation of solar energy technology. This system uses hundreds, or even thousands, of flat mirrors called heliostats. The heliostats track the sun, concentrating its energy up to 1,500 times its natural intensity. This energy is transferred not through a heat-transport fluid but by reflection from the heliostats directly to a single receiver—a heat exchanger mounted on a central tower. By transferring the sun's energy via reflection rather than through a heat-transfer fluid, solar power towers reduce the amount of heat lost on the journey through the pipelines of a parabolic trough or solar dish system. In addition to improving the efficiency of solar heat collection, solar power towers can also reduce the costs of installing a solar field, as the flat mirrors are much less expensive to produce than concave ones.

The rub is that, for a solar power tower—or any solar operation really—to be economical, it's got to be big, and big has inherent expenses.

Let's talk a little about costs. The price to build any energy-generating system is calculated as the cost to install 1 kilowatt-hour of generating capacity, and right now the cost to install 1 kilowatt-hour of utility-scale solar power is approximately $2,000 to $3,000. This figure includes the cost of the land that must be purchased—and, with current technologies, the amount of land required to install a solar field is about 5 to 10 acres for each megawatt of electricity.

Whether a solar plant can be a practical part of your town's energy plan depends on the amount of land your community can dedicate to a solar field. Florida, for example, is nicknamed the Sunshine State, but the state's geography, with its large urban centers, inland agriculture, extensive shorelines, and environmentally sensitive wetlands, makes it a poor candidate for solar development. In the United States, in fact, current technology really only allows utility scale solar to work well right now in the southwest, though thin film PV may make solar a practical option for a wider range of locations within the next several years. Thin film solar, indeed, offers us the biggest opportunity to ramp down the price of technology faster than any other renewable, and thus lower the cost of the associated electricity.

Then again, you could choose to purchase a portion of your town's electrical power from a solar plant that's not necessarily located nearby. Whether doing this makes sense from an economic perspective—right now the cost of solar-generated electricity can range from between 9 and 16 cents per kilowatt-hour—depends on many factors, including the portion of power your state, regional, or federal government mandates that you purchase in the form of renewable energies, the quality of the incentive programs available to you to fulfill your mandate (which we'll discuss in more detail in chapter 7), and how the grid system is improved so that the electricity can be delivered to you with ease and reliability. We'll get to *that* discussion in chapter 5; for now, let's just acknowledge that solar development is hampered by the fact that getting the sun-generated electricity to customers is not always practical given the shortage of transition lines and capacity constraints on our current grid system.

But we don't want to make it seem as if incorporating solar power into your overall energy plan is knotty; just the opposite, we want to encourage you to think out of the box. For example, one essential for generating economical solar power is an expanse of flat, black surface. Dr. Arian de Bondt, an engineer with the Dutch building company Ooms, looked around one day at Scharwoude, the city where his company is headquartered, and imagined what could happen if, instead of building special, dedicated solar surfaces, we used the flat, black surfaces we already have in abundance: asphalted roads.

Dr. de Bondt devised a system of heating and cooling that employs Scharwoude's roads as a sort of a solar field. His system uses a circuit of connected water pipes running from one side of the street to the other, most just under the asphalt layer. When the surface of the street is hot, water pumped through these pipes picks up the heat and transports it to a natural aquifer, to a series of heat exchangers. The street-heated water warms the groundwater before returning to the surface by way of exit pipes, turning the aquifer into a heat store. In winter, the heat that was stored during the summer months is pumped into the Ooms building to keep it warm. On its way back out of the Ooms headquarters, back through the pipes under the streets, the water is then cooled by the freezing temperatures of the winter asphalt and is diverted into a second aquifer, where the cold water is stored until it's needed in the summer, to cool the building. The result is that the Ooms building enjoys heat in the winter and air conditioning in the summer, without the use of one lump of coal or one drop of fossil fuel oil.

Oh, and one more thing. The summer sun naturally softens asphalt, making it vulnerable to damage from traffic. One of the residual benefits of Dr. de Bondt's system has been that the winter-cold water circulating below the road surfaces in summertime, on its way to cool the Ooms building, keeps the asphalt hard and saves the city the expenses of costly road repairs.

Try to wrap your imagination around that: Road damage normally caused by the heat of the summer sun is prevented by using the power of the sun itself. As technologies like Dr. de Bondt's are developed and expanded, some people just might be tempted to assign the unexpected benefits of renewable energy to having "fooled Mother Nature." We prefer to think of them as evidence of the value of acting in partnership with her.

Wind Power

If you're a sailor—or, like us, lucky to have friends who are sailors and will invite you along for a weekend on the water—you already know the satisfaction of harvesting the wind, the accompanying fullness of spirit when the sails are billowed taut and you are skimming over the water at 10 knots under the power of a force of nature that's been tamed only with your wits and a bit of cloth.

This is an ancient satisfaction. The wind was first harvested by sailors who couldn't have explained the principles of aerodynamics to you on a bet but who nonetheless knew how to rig their boats and trim their sails in the service of trade and exploration. Wind energy has been used to power transportation since the time of our earliest ancestors— and it's been used to power industry for hundreds of years. It may be, Mr. or Madam Mayor, that wind is an appropriate part of the renewable energy mix for your hometown today.

Windmills can be found all over the world. The first, it seems likely, came from Persia, with the technology spreading to northern Europe as a result of the Crusades, though the earliest documentation of a working windmill is from 1219 CE, in China. These early windmills were used to pump water and mill grain. The Dutch, as early as the 1300s, were refining the windmill's design from simple upright wooden "post" mills into the sturdy, multistory "tower" mills of brick and stone that are one of that country's tourist touchstones today. In these tower mills, separate floors could be given over to grinding grain, removing chaff, and storage as well as to living quarters for the "windsmith" and his family. The windsmith's job was to optimize the windmill's energy output by manually orienting its rotors to the wind and to furl those rotors during a storm to protect them from damage. Through research and development—thoughtful trial and error—the Dutch adapted the technology of windmills to such a sophisticated degree that their Industrial Revolution was powered, in large part, by the wind.

Windmills are nothing new to North America either. The blustery plains of the American Midwest, for example, provided prime conditions for harvesting wind energy, and the pioneers moving westward dotted the landscape with mills given over to the same traditional purposes, pumping water for agriculture and milling grain.

In 1888 in Cleveland, Ohio, a windmill was used for the first time to generate electricity. In the years that followed, America's farmers increasingly connected their windmills to generators and used the power to feed electric lights and to charge their crystal radio sets. It was, in fact, the relative luxuries of light and radio accessed by these early electric windmills that helped to drive the demand for power in the heartland. As people in rural communities began to acquire the newfangled appliances that were starting to be commonplace in the cities—electric vacuum cleaners and electric frying pans and electric sewing machines—they became dissatisfied with the intermittent, weather-dependent electricity their mills could generate. They were thrilled by FDR's electrification program.

Harvesting wind energy became less of a quest in the decades that followed electrification, but the development of wind power technology was never wholly abandoned. In 1973, during the Arab oil embargo, for example, the U.S. federal government turned its attention earnestly to wind. The federal wind energy development program, however, proceeded in fits and starts, surges of interest that corresponded tellingly to fluctuations in the price of oil—and to fluctuating commitment levels by members of Congress impatient to see results and ever ready to pull funding if they decided the technology wasn't improving quickly enough.

Still, by 1980, 8 megawatts of electric energy were being generated by wind farms annually in North America—the same year in which all of Europe, which seemed to have had both a head start and a deeper commitment to wind power, generated only 4 megawatts. Fifteen years later, in 1995, the balance had shifted—North America was producing 1,678 megawatts of wind-generated electricity to Europe's nearly 3,500 megawatts.

Today wind power is the fastest-growing segment of the renewable energy industry. More than 90,000 wind turbines generate over 7.8 billion kilowatt-hours of electricity annually worldwide. Wind energy accounts for nearly one-third of all renewable energy installations, and a full two-thirds of this wind power is installed in Europe, where Germany, Spain, and Denmark are the top wind-power producers. In Denmark, the clear leader in the wind-power sweepstakes, 18 percent of the electricity used in the nation comes from its turbines. The United States and India round out the ranks of the top five wind-energy users, a quintet of countries that accounts for 70 percent of all of the wind power used in the world. India today, in fact, is experiencing a conscious energy evolution of the sort the Rural Electrification Administration achieved in America; the national slogan for this electrification program is "Power for all by 2012." The Indian federal government has introduced legislation that will require a specified amount of its growing energy needs be met with power from renewable sources, and right now wind energy is the option its people are choosing to provide their power. The elusive wind isn't so elusive these days; it's an $11 billion global industry projected to grow by 27 percent a year.

The most fundamental explanation for how wind energy works is that it's the exact opposite of a normal household fan. A fan uses electricity to generate wind; a turbine's rotors are activated by the wind, which transfers this natural power to a generator to produce electricity that is then fed into the existing grid system. What are the costs of accessing this elegantly elementary process?

Like solar, wind energy might seem as it if ought to be pretty cheap. Once installation costs are covered, the fuel, after all, is free, and there is an infinite supply of it. But also like solar fields, wind farms are land-intensive. Specifically relative to wind, this cost includes the often-rented expanses of land on which the turbines are placed. An average turbine for small or private wind-power generation is thirty feet high but can be as tall as twenty stories for a commercial installation. An installation that produces 1 megawatt of wind-generated electricity requires fifteen to twenty acres of land, and landowners frequently command up to $3,000 annually per turbine; this can be a boon to farmers in creating uses for land that might otherwise be unsuitable to agricultural purposes, but it does raise the cost of the wind.

The cost, on average, to install 1 kilowatt of utility-scale wind power is $1,600. This might seem like a bargain in comparison to coal-produced electricity, which costs, on average, $2,200 per kilowatt to install, or nuclear, which costs $4,000 per kilowatt. And it might seem excessive compared with the $1,100 per kilowatt-hour required for natural gas. But in developing a comprehensive cost-benefit ratio for the renewable power sources you're considering for your town, the calculations aren't so cut and dried. Beyond the installation costs, you have to consider reduced CO_2 emissions, decreased human health care costs, and increased energy security—all of the upsides that come with renewable energy—as well as the downsides that might be particular to each sort of renewable source. In the case of wind, the particular downside is that though the cost to install 1 kilowatt-hour of electrical capacity is about $600 less than to install a coal-fired plant, wind farms are less efficient energy generators, generally by about 30 percent.

But bringing the cost of wind power to a more personal level, let's say that a private homeowner in your city has a property located in an area with conducive wind conditions like a flat plain, or a high mountain, or a coastal region. If he installs one wind turbine with a capacity to generate five to fifteen kilowatts of electricity to augment the power he receives through the existing grid, he could save as much as 50 to 90 percent on his electric bill—and that's a cost-benefit ratio well worth looking into.

So far we've been talking almost exclusively about land-based wind farms, which utilize mature technology that employs generally standardized equipment. But as we pointed out, coastal regions are one of the prime locations for wind farms, and, more and more, engineers are looking to plant wind turbines offshore. Given the great velocity of ocean winds, it makes sense to consider the use of offshore wind power if your community is located near a coastline. The higher-velocity offshore winds increase a turbine's generating capacity and improve its efficacy as an electricity producer.

But offshore wind-power technology isn't yet fully mature, and a lot of questions need to be answered before it has a real chance of becoming conventional. For example, how will the equipment endure in the tough offshore weather conditions—subjected to those high wind speeds? The farther a wind turbine is located from the shore, the longer the distance the cable must reach to connect to the grid—and each extra mile increases cost. How does that added distance figure into the cost of sinking the cable versus the value of the power the cable will be able to deliver? And how much electrical loss will be sustained in the energy's journey over the longer distance? Does the loss negate the benefit? And what of the potential for environmental impact?

In addition to these critical questions, there have been objections to offshore wind farms from the coast of Massachusetts to the shores of the Netherlands because, frankly, a set of giant turbines off in the distance doesn't do much to enhance the ocean view. We'd have to agree that preserving the aesthetic integrity of the environment is an important part of environmental assessment. In response to the aesthetic objections, some potential offshore–wind farmers have proposed moving their projects even farther offshore, so the turbines won't interfere with coastal vistas. This solution, once again, presents a positive and a negative that have to be carefully balanced: The velocity of the wind increases the farther offshore you go, so the additional distance might just increase turbine efficiency to some startlingly significant degree. But those same even higher velocities could exacerbate the problem of potential equipment failure and will increase the costs of connection to the grid, so the answers to the questions we must pose about offshore wind power become just that much more difficult to come by.

Still, given all of the benefits of clean, wind-generated energy, what else could stand in the way of realizing the million-plus gigawatts of land-based capacity that's estimated is yet to be developed? What are the other objections you, as a mayor, might face from your constituents if you propose to build a wind farm to help power your town?

Well, noise. As wind-power technology has been developed, turbines have acquired a reputation for making a lot of it. That reputation is now undeserved. Engineering breakthroughs have reduced the noise levels of turbines to the extent that from a distance as near as 400 yards, the sound of a wind farm is no more than you'd hear from a normal household air conditioner. From a distance of only one mile, there is simply silence.

Concern about wildlife—birds, in particular—is another issue some people have with wind power. Early windmills could be hazards for flocks of various migrating species including golden eagles, red-tailed hawks, owls and kestrels, all of which are protected in the United States under the Migratory Bird Treaty Act or other codes. Since the 1980s windmill engineers and wind farmers have been consulting with conservation and environmental groups to adapt the technology. The result is that they've altered the design and speed of contemporary turbines to make them safer for birds, and adjusted the placement of the windmills themselves so as not to interfere with established migratory patterns. The simple addition of strobes on the rotors is another safety feature that warns and repels flocks.

It's aesthetic objections, however, that seem to be the sticking point for so many who face the possibility of a wind farm on or near their property. Contemporary windmills admittedly lack the quaint charm of their early Dutch ancestors. When we're planning a wind farm—or when we're planning an installation of any other sort of power plant, for that matter—it makes sense to be sensitive to the concerns of the local community for whose benefit we're building it.

On a recent trip to California, traveling by car from San Francisco south to Santa Barbara, the giant turbines rotating on expanses of otherwise undevelopable land were awe-inspiring—a sight that, when viewed with an understanding of the enormous contribution they were making to our energy security, was truly one of great beauty.

Hydroelectric Power

Like the sun and the wind, water has been serving the industry of humankind for at least as long as we've been recording history. In Himalayan villages in the twelfth century, for instance, small hydro systems powering waterwheels to grind grain would have been a common feature of the landscape. The first hydro*electric* power plant was built in Appleton, Wisconsin, in 1882. It generated just 12.5 kilowatts of power, enough to provide lights for two small paper mills and one house. Today hydroelectric power is a form of renewable energy so mainstream that often it's not even included in state or national tallies of renewable power capacity.

One of the reasons why hydroelectric power has such a comfortable niche in our contemporary energy mix—and in our perceptions—is the lockstep progress of hydro technology and the electrification of developed countries. It's been around for a long time. Remember that it was FDR, the father of American electrification, who called for hydroelectric power to be expanded in the 1930s as a source of cheap, clean electricity.

Nowadays hydroelectric giants, such as the 7,600-megawatt Grand Coulee power station on the Columbia River in the state of Washington and the 13,000-megawatt Itaipu Dam on the Paraná River in Brazil, combine to provide 24 percent of the world's electricity. The global capacity from all hydroelectric facilities is 675,000 megawatts producing 2.9 trillion kilowatt-hours of power annually and—not incidentally—saving the world the equivalent of 1.7 billion barrels of oil each and every year.

Hydroelectric power works by converting the energy in flowing water to electric energy. The same physics lie behind the design of all hydroelectric systems: A dam is used to capture and store water; pipes, or penstocks, carry the water from a high reservoir, downhill, toward turbines in a power station, and the strength of the natural pressure of the surging water is often increased by nozzles affixed to the end of the pipes; the water strikes the turbines, rotating them and driving a generator that produces electricity.

But, though the typical arrangement of all hydroelectric stations is similar, there are several variations in how the design can be implemented. First of all, the amount of electricity any hydro system can produce is determined by its "head"—the difference in height between the surface of the water in the high reservoir and how far it must fall to reach the turbines in the power station below. The greater the height, the more power that is achievable. The height of a system's head is determined as much by engineering feats as it is by the natural geological features of the chosen site.

Second, the water's flow can be utilized in a variety of ways. *Conventional* hydroelectric power plants use a one-way flow of water. These sometimes are called run-of-the-river plants, meaning that they rely solely on the native flow of the body of water. Consequently, they are significantly affected by the weather and by seasonal changes in rainfall and water levels that can cause fluctuations in the amount of electricity they produce. Conventional hydroelectric systems, with one-way water flow, can also be designed as "storage" plants, which reserve enough water in their dams to offset seasonal impact on their water flow. Large dams can, in fact, store several years' worth of water.

Other hydroelectric systems are designed as *pumped storage plants*. This means that after the naturally flowing water has produced an initial quantity of electricity, it's diverted from the turbines into a lower reservoir below the dam. During off-peak hours, or through dry-weather conditions, the water in this lower reservoir can be pumped back up and reused to supply a steady stream of electricity to the plant's customers during peak use times. This characteristic of water—that it can be stored in dams or redirected for reuse—has probably also given hydroelectric power a historic advantage over the intermittent wind and sun and has likely contributed to its rapid acceptance and use as a mainstream power source.

The sort of hydroelectric system that most of us are familiar with are the colossal plants like the Grand Coulee and Itaipu dams. Most of these monster systems have been built by government agencies, and their purposes extend even beyond electricity generation to flood control, irrigation, and recreation. But although nearly a quarter of the world's electric energy comes from this clean and renewable resource, it's unlikely that many new hydro systems will be built on a grand scale. The costs of such massive construction, coupled with environmental concerns, limit projects of this scope.

So, then, where does the future of hydroelectric power lie? The world needs more and more electricity, and it's estimated that over 78 percent of this resource's global potential remains to be tapped. How do we go about tapping it? Well, new hydroelectric facilities don't necessarily have to be built big in order to serve our needs well in the coming renewable age.

One of the most common objections to the development and construction of new monster hydroelectric plants has been their impact on the surrounding ecosystems. In truth, farmlands and wildlife habitats are indeed affected by such hydro systems, and the efficacy of mitigation measures like fish ladders for spawning salmon can make for some heated conversations.

But a country's interior villages and cities usually were founded on waterways—navigable rivers that gave the settlers access to transportation as well as water to power their mills, to drink and bathe in, and from which to harvest fish to eat. The flow of these rivers is often already controlled by dams for the purposes of flood control or irrigation.

In the United States alone, for example, there are 80,000 existing dams, and only 2,400 of them are currently being used to generate hydroelectricity. What if these existing dams were retrofitted for hydroelectric capabilities? Is there a dam in your town, Madam Mayor? Is it suitable for retrofitted operation? Adapting even a portion of the dams America already has to make them hydroelectric producers could substantially boost the percentage of power the country generates from this renewable resource—and at the same time as it neatly avoids the controversy that often accompanies new construction.

Many such smaller-scale hydroelectric facilities are, in fact, in use today. In the 1980s, for instance, FDR's New York Power Authority built five smaller hydroelectric projects in rivers and existing reservoirs across the state specifically as a way to head off the state's increasing dependence on burning foreign oil to produce its electric power. The smaller scale of these projects has allowed them to be implemented without affecting the quality of the water used to run the systems or disturbing the conventional uses of the waterways, such as recreation and navigation. With a combined power of nearly 36,750 kilowatts, however, they have achieved their purpose of helping to free the state from total reliance on imported oil.

Because of the vast and worldwide opportunities for the expansion of hydroelectric power—and because hydroelectric power is clean and cheap; the cost to install 1 kilowatt-hour of hydroelectric power is about $2,500—hydroelectricity is a natural in every sense of the word.

Before we finish with hydroelectric systems, there is one more—and even more localized—way to employ water power to produce the energy we need. "Micro-hydro systems" are the smallest and, so far, least common of all hydroelectric-generating systems. Generally defined, they are any hydroelectric system with a generating capacity of less than 300 kilowatts, and they're suitable for installation anywhere there is a swift, steady flow of water and/or in areas where there is a consistently heavy rainfall.

Just how "micro" are we talking about? There are designs for hydroelectric systems that, with less than three feet of head, can produce enough power to run two or more houses, and the equipment for such a small system can usually be had for less than $1,000. Micro-hydro systems can work for homeowners with even a small creek on their property—lowering a monthly electric bill, feeding any excess of electricity back into the grid, and contributing to the reduction of CO_2 in the atmosphere.

Micro-hydro systems also hold great promise for bringing electricity to remote areas of developing nations where more traditional systems would damage fragile or pristine ecosystems, or to which it would be too cumbersome and costly to transport traditional fuels. They can provide clean power to people who are at present denied this modern necessity. With it, they can provide people the means to—as President Eisenhower phrased it so succinctly—improve their lot. In short, often it's not the scale of the power plant that's important, but the incremental contribution each individual system, colossal or micro or any size in between, can make to lessen the burden of our energy predicament, all systems working in concert to improve the lot of the world.

Geothermal Power

Geothermal energy might well be the least understood of all of the types of renewable resources we've talked about so far. Certainly it's the least visible. The word "geothermal" comes from the Greek *geo*, meaning "earth," and *therme*, meaning "heat." Geothermal energy systems make use of the heat that's produced deep inside of the earth, at its core, 4,000 miles below the planet's surface.

The earth's core actually has two layers: an inner core of iron and an outer layer of very hot melted rock, called magma. On top of the magma layer comes the mantle, a layer of earth that's about 1,800 miles thick, made of magma and rock. On top of the mantle is the earth's crust, a layer that's relatively thin—from three to five miles deep beneath the ocean and fifteen to thirty-five miles deep under landmasses. The earth's crust is broken into pieces called plates, and it is at these broken places that heat manufactured in the magma layer by slowly decaying radioactive particles comes closest to the surface. The heat manifests itself as volcanoes, geysers, and hot springs—places where the Romans, Chinese, and Native Americans built their baths, and where today spas still draw enthusiastic patrons to their mineral-rich healing waters. It's these same places that are, in general, the best locations for geothermal heating and electric-generating facilities.

All geothermal systems rely on two basic components: the heat beneath the earth's crust and the subterranean waters that the earth's heat will turn to steam. In most geothermal systems, accessing these components involves drilling up to two miles into the earth's crust. In direct heating systems, the earth's natural steam is piped directly into buildings to warm them in winter and—perhaps surprisingly—to cool them in summer.

How does *that* work? While the seasons change from cold to hot and back again out here on the surface of the planet, the temperature in the upper ten feet of the earth remains fairly constant, at between 50° and 60° F. The benefits of this constant temperature can be accessed by pumping the water of springs or reservoirs near the earth's surface into buildings for interior climate control. In some cities in Iceland, a leader in using geothermal technology, the climate in nearly 95 percent of its buildings is managed in this manner.

Geothermal power can also be used to make electricity, of course; it already supplies over twenty countries, including France, New Zealand, Russia, China, and the Philippines, with about 8 percent of the renewable energy generated globally. Though at the present time it costs between $4,000 and $5,000 to install 1 kilowatt, it has the potential to become a very cost-effective way to produce electricity, and its development potential is broad worldwide, so the technology deserves to be a little better understood.

There are four different ways to drive electric generators using geothermal energy. The first is called the dry steam method. First developed in 1904 by Prince Piero Ginori Conti at the Lardello field in Tuscany, this method uses the steam released directly from a geothermal reservoir to drive generator turbines.

A more technologically sophisticated method of geothermal electrical generation is called the flash steam system. This is the most common system in use today, and it works by taking advantage of the high pressure beneath the earth's crust. Under this intense pressure, water remains liquid though it's heated to what would be well over the boiling point were it at sea level. As the water is pumped from within the earth, an abrupt drop in pressure causes it to convert—in a *flash*—to steam, which more efficiently powers the turbines that energize the electrical generators.

Most geothermal facilities that are now in the planning stages incorporate a third, and even more efficient, technology to access geothermal power. Called the binary system, this method directs the earth's hot water to a heat exchanger, where the heat is transferred to a second pipe containing a fluid with a much lower boiling point than water, usually either isobutane or isopentane gas, which is then vaporized to power the turbines. The advantage of this system is that it can make use of those geothermal reservoirs that have lower temperatures, which increases the places where geothermal systems can be located.

Finally, enhanced geothermal, or the hot dry rock system, may be yet another avenue into deep earth's power potential. Rather than harvesting the heated water already in the earth, this method involves manufacturing steam by piping surface water into the hot but dry rocks in the earth's crust. The benefit of this system is that it can be used anywhere on the planet simply by drilling a hole. The downside is that the hole has got to be dug deep—deeper than for any other geothermal system—and the environmental impacts of deep drilling aren't yet fully understood.

The United States is the leading producer of geothermal energy, with plants in Nevada, Hawaii, Utah, and thirty-three facilities in California that alone produce about 90 percent of the nation's geothermal electricity, with another 35 megawatts of capacity planned in 2008. What are the downsides of this rapidly expanding technology?

Well, for one thing, geothermal facilities do release CO_2 and other pollutants, such as sulfur, that might be contained in the steam. As we look to new sources of electricity that reduce greenhouse gases, however, geothermal remains one of our best opportunities; the emissions are only 1 to 3 percent of that generated at one of today's standard fossil fuel–fired plants. In addition, this percentage can be lowered even further by incorporating "clean coal" technologies, such as scrubbers to remove sulfur, at geothermal plants.

Additionally, geothermal plants need to be carefully managed. Geothermal energy is renewable in the sense that the water beneath the earth's surface constantly replenishes itself. Even after it's been drained, a subterranean reservoir will refill itself, but the process can take years if the natural supplies aren't managed judiciously. In every method of geothermal energy generation, unused steam is injected back into the earth, recycling it. In many systems, the water in a geothermal reservoir is replenished with gray water—treated sewage water that would otherwise be piped into streams and rivers. The trick is to replenish the water in amounts and at just the right times so that the earth's heat isn't cooled by the supplemental water.

But these are all management issues well worth working around. A 2006 report by MIT scientists, which takes into account the use of enhanced geothermal, estimates that there is enough energy in rock just about five miles beneath the United States alone to supply all the world's electrical needs at the current rate of demand for 30,000 years. Moreover, there seems to be no reason why the steam from a geothermal plant couldn't, with the development of the appropriate technology, be used to feed an existing coal-, oil-, or even nuclear-fired generating plant. With potential like that, we have got to pay attention to what geothermal has to offer.

SAVING SUNSHINE FOR A RAINY DAY

We need to thank our nine-year-old daughter, Katie, for helping us to write the next few paragraphs. Over dinner one evening, Michael and I were talking about a way to explain why the world's energy plan needs to be integrated, to encompass several different renewable technologies that generate electricity in order to be successful—why we need to build solar fields *and* wind farms *and* hydroelectric facilities *and* geothermal plants in order to secure our energy future.

Katie chimed in. "Why don't you just say we need solar fields because the wind isn't always blowing and we need wind farms because the sun doesn't shine at night?"

Out of the mouths of babes.

Grown-ups really do overthink things sometimes.

Keep in mind that what we have been talking about are technologies that harvest the magnificent powers of the earth for human benefit. And the earth has rules about that. For instance, as we've noted, each sort of renewable energy might work optimally at a different time of the day (the sun), or within particular weather patterns (the wind), or because of specific natural features of the landscape where the individual facility is located (hydroelectric and geothermal). We can't very well "stoke" the wind, as we might a coal-fired power plant, revving it up to maximum capacity to produce its power whenever it suits us. Nature, though, serves us well in this regard: The sun conforms to the same business hours that most of us keep and is therefore available as a source of power during peak times of the day, when our need for energy is greatest.

One of the major criticisms leveled at renewable energy, however, especially at solar- and wind-generated power, is that we don't know how to store the electricity they create. Therefore, the sun and the wind aren't very reliable sources of power for our fast-paced, digital world. But let's face it: We don't know how to efficiently store *any* electricity made from *any* fuel source. Like us, you're probably more than familiar with the frustration of a child whose favorite toy has run out of batteries. Today batteries are indeed the most convenient way to store electrical power. But employing batteries on a scale that would power towns and cites, school districts and industries, would, with current technologies, be wildly impractical. Storing electricity is a most cutting-edge technological challenge because, in the whole history of human reliance on electrical power, we've never really had to do it before. We've just thrown more fossil fuel into the burner at the power plant. Storing electricity becomes a challenge that is specific to making a more seamless transition to a 100 percent renewably fueled world.

Here are some things we can do to circumvent the storage problem as we transition into the new energy age. The first step is to use renewable energy sources in a complementary strategy, each one backing up the other, in the places that are best suited to the installation of each sort so that our homes and businesses always have a steady supply of clean electricity.

As technology progresses, however, and we develop ways to store renewable power efficiently, we'll improve the consistency we can enjoy from wind and solar energies. Engineers are even now working on ways to do this.

Some solar energy providers, for example, are using thermal-clad tanks to store sun-heated water. This allows their systems to run without sunlight for half an hour or so—long enough a for solar energy producer to avoid incurring penalties for failing to ship to the grid the full amount of energy that it has contracted with a distributor to supply. Other solar energy providers opt to store the sun's energy in a salt system. The high temperatures that can be sustained when holding sun-heated water in molten salt allows these systems to continue to produce electricity for up to sixteen hours in the absence of direct sunlight— overnight, or on a cloudy day.

Still other solar fields are hybrid solar facilities, meaning that when the sun isn't shining, they have the capacity to switch to another fuel source—perhaps a low-carbon fuel like natural gas—so that they can continue to generate electricity around the clock or in inclement weather.

One developing technology for generating solar power neatly sidesteps the storage problem altogether. *Solar chimneys* are different from other ways of generating solar power that we've mentioned so far in that they don't use a reflective surface, such as a mirror, to focus the sun's energy. Instead, solar chimneys work by capturing the sun's power in enormous "greenhouses." As the air in the greenhouse heats, it rises to the center of the structure and up a long chimney that's lined with turbines. Because the greenhouses are so vast, they can store enough hot air to keep the turbines rotating twenty-four hours a day. Right now an Australian firm, Enviromission, is working with the German consultants Schlaich Bergmann to build five 200-megawatt solar chimney systems in New South Wales. The company plans to have these experimental chimneys—each of which is almost 1,100 yards high and four and a half miles wide—operational sometime in 2008.

As to catching and storing the wind—which, as Katie points out so pithily, simply isn't always blowing—there are pioneering experiments afoot here as well. One of them is being developed in Vancouver, British Columbia, by a company called Encore Clean Energy, Inc. Encore's technology is based on the compressed air energy storage (CAES) systems that typically have depended on fossil fuel–burning power plants to compress the air they use. This air is then stored in underground caverns and used to produce electricity at peak hours, when we demand it most.

Encore is working on a system that could allow wind farms to store energy in the form of compressed air in underground tanks or pipes and release it through a special generator to create electricity when the wind is still. The special generator is the key piece of equipment; it's called a magnetic piston generator (MPG). The MPG is essentially an internal combustion engine, but, through a unique arrangement of the pistons, it can be driven with air pressure to generate electricity cleanly, with far less emissions, than conventional internal combustion engines.

A company called General Compression, out of Massachusetts, is also working on a compressed air wind storage system. Named the Dispatchable Wind Power System (DWPS), their product is a comprehensive wind farm/storage installation comprised of special turbines equipped with the company's trademarked Dragonfly compressors, storage units, and expansion/generation equipment. The DWPS can be installed in existing wind farms to double or even triple a facility's current power production.

One drawback to compressed air storage is: Where to store the compressed air? General Compression recommends its system for locations that have natural geographical features suitable for storage purposes. This form of storage isn't necessary to the installation of a successful system, but it is an asset that can greatly reduce the cost of a wind storage operation—and reducing the cost of renewable sources of energy is always a primary goal; bringing the price tag of renewable energy in line with fossil fuel-generated power is what is going to make clean energy conventional.

BIOMASS

Beyond developing the technologies that will allow us to store solar and wind energy as effectively as we can store water in a dam or the earth stores heat in its core, we can also look at using innovative materials to feed the burners in the plants that generate our electricity. Trees, grasses, agricultural crops, and other biological materials are collectively known as biomass. Many people probably associate biomass with the manufacture of alternative fuels—ethanol and biodiesel—and well they should. We'll talk in depth about how agricultural products such as corn and sugar as well as nonfood crops like switchgrass are used to make these fuels in chapter 6. Here we're talking about how wood waste, biogases, and even the scraps in your garbage—yard waste and paper that can't be recycled into new paper products—potentially can be used as fuel in power plants rather than taking up space in a landfill. In the southeastern United States, as a matter of fact, biomass technology is already leading the region's renewable power potential.

Wood is the most common form of biomass. In the United States, about 2 percent of the energy manufactured today comes from wood and wood waste, such as bark, sawdust, wood chips, and scraps, much of it from industries that use wood as a raw material and recycle the scrap to create their own energy supply. In landfills, when biomass rots, it produces methane, as does the manure at dairy and poultry farms; this gas can be collected and processed in tanks called digesters to produce power. Even your trash—known more formally as municipal solid waste (MSW)—contains food scraps, leaves, and lawn clippings that can become feedstock for power plants. But how much biomass can your lawn clippings and such really amount to? Well, right now, for example, the state of California produces more than 60 million tons of biomass each year. Less than 10 percent of that total is burned to make electricity, but if all 60 million tons were used, it could generate nearly 2,000 megawatts of electricity, enough to power 2 million homes!

Some studies estimate that in the entire United States—in just this one country alone—there is an available biomass of 1.3 billion tons per year. But let's take the question of biomass down to an even more gut level: 35 percent of the food purchased in Britain, and 50 percent in the United States, ends up rotting in a landfill, producing methane that contributes to global warming but that might be used for more constructive purposes.

Burning biomass, however, is an art, not a science. At the very least, it's a more thoughtful process than simply shoveling in a more traditionally consistent fuel, such as coal. The amount of moisture in the yard waste, or whether the wood waste is in the form of sawdust or wood chips, all require that the burners be adjusted or even adapted to handle each specific type of feedstock.

Additionally, biomass—because it's composed of decomposing vegetation—contains carbon that it will release when it's burned. But because the tree in your backyard, for instance, produces new carbon-eating leaves every year to replace the ones you've raked up and sent to the power plant, the level of carbon in the atmosphere remains relatively stable when biomass rather than, say, coal is burned as a fuel.

As storage techniques for renewable energy sources continue to be developed and as biomass comes into more common use as a feedstock for electric plants, other creative solutions along wholly different lines of thinking present themselves as well. Case in point: strategic solar reserves.

Most readers know that since the oil embargo of the 1970s, the U.S. federal government has kept on hand a Strategic Petroleum Reserve (SPR). The SPR is the largest stockpile of government-owned oil in the world. Nearly 1 billion barrels of crude oil are now on hand that can be used in cases of national emergencies, as the SPR was during Hurricane Katrina in 2005.

Along the same lines as the SPR, in 2007 the U.S. Congress's National Resources Committee called for public lands that have "high solar incidence"—plenty of sunshine—to be identified and set aside as strategic solar reserves. These tracts of land would be developed as solar fields to generate the energy the nation needs in cases of natural disasters and other national emergencies.

You might wonder at the idea of solar energy as an emergency power source, especially in light of the storage issues we've just talked about. But emergency solar energy is a concept that could have been directly inspired by the recent calls for renewable energy sources from military commanders currently stationed in Iraq—people who know from *emergency*.

In September 2006 Marine Corps Major General Richard Zilmer sent the Pentagon a "Priority 1" request for solar panels and wind turbines, as a "self-sustaining" solution to the military's energy problems in oil-rich Iraq. Oil to generate the electricity the military needs to carry out its mission in that country might be *produced* locally, but it still has to be *transported* for long distances from its source to the soldiers. Convoys hauling fuel to U.S. outposts in Iraq are moving targets, vulnerable to ambush by insurgents, and driving a fuel truck is one of the most dangerous assignments of the war. The need for oil to run military operations, according to Zilmer's memo, leaves troops "unnecessarily exposed," and, without renewable power, they will "continue to accrue preventable . . . serious and grave casualties." In addition, the hot "thermal signature" of diesel generators currently in use calls enemy attention to troop locations. So commanders are urging the Pentagon to supply them with the equipment they need to generate renewable electricity and liberate their operations from dependence on oil.

Better than any of the rest of us can possibly imagine, the military understands the wisdom and necessity of the strategic freedom that comes with being able to rely on truly independent sources of power.

Each sort of renewable energy we've talked about in this chapter—solar power, wind power, hydroelectric power, geothermal power, and power from biomass—is a distinct resource, and we've provided a separate description of each source. We'll stress again, however, that the hallmark of a successful renewable energy plan will be to value all the sources, each one as an essential part of an integrated system. As you look around your city from your lofty vantage point as its mayor, searching for its best opportunities to employ renewable energy technologies, envision the sun and the wind, the water and the earth as one harmonious whole.

All right. Now we've generated all of this wholesome, renewable electricity. How do we get it to the people who need it? Transporting electricity to its end users is the next part of our energy challenge.

5

POWER TO THE PEOPLE

A MAN SLIPS ON A BANANA PEEL . . .

A MAN SLIPS ON A BANANA PEEL. AT THE END OF HIS FALL, THE MAN'S HEAD comes down on the tines of a rake so that its handle flies into the air and connects with the back end of a sleeping cat. The startled cat shrieks and jumps off the ledge, where it had heretofore been curled up so peacefully, onto one plate of a counterbalanced scale so the opposite plate swings suddenly upward, bringing with it the match affixed cleverly to its side, which strikes against a cunningly positioned emery board along the way, which ignites the pilot on the nearby gas stove so the water in the kettle on one of its burners comes to a boil so that someone can have a cup of tea.

We always hope it's the same man who went to all the trouble and discomfort of slipping on the banana peel in the first place.

A "Rube Goldberg" is an invention that takes a series of complicated, comic steps to perform one simple, everyday task. People have been laughing at Reuben Goldberg's drawings of these marvelous contraptions for a hundred years, but most of us would be as startled as that poor sleeping cat to realize how much our contemporary electrical distribution system has in common with the comics, albeit with little of Goldberg's whimsy.

Very few of us think much about what happens after we plug our coffeemaker or our toaster into a wall socket. And why should we? All we really want is breakfast, and that outlet was designed so that we can prepare the meal by using the ultimate in labor-saving conveniences: electricity. Since wall sockets started to become common in the 1930s, we've gotten so used to the convenience that there's been little incentive for us to learn about how electricity is made, traded, moved around, or delivered to us on demand.

But the growth in worldwide demand for electricity has been nothing short of radical. And changes in the laws that regulate how electricity is delivered in the developed world, as well as the ways in which new delivery systems are being structured in developing countries, should incite us to pay a little more attention. How power is delivered to our homes and businesses is another part of our energy present we should understand in order to see more clearly the benefits of the renewable resources that are our energy future. Electricity delivery infrastructure is—or ought to be—a puzzle that keeps system engineers, suppliers, and lawmakers at the federal level of every country on earth awake at night. Any substantial interruption in electrical service could shut down the economy and jeopardize human welfare in ways we don't want to imagine—food supply chains, sanitation systems, communications systems are all dependent on a power distribution infrastructure that is desperately stressed and in critical need of creative upgrading.

Reimagined electrical delivery infrastructures are, fortunately, a part of our energy future wherein municipal officials, such as mayors, can have a significant and positive influence.

Because the centralized electricity delivery system devised in the 1930s in the United States was once considered world class—the gold standard of delivery systems, if you will—it can serve as a template for how existing grids can be restructured, and how emerging grids can be designed, so the power keeps flowing to us seamlessly, without interruption, just in the way we are accustomed to it behaving.

In the United States today, there are more than 3,100 electric utilities operating over 10,000 power plants. Of these utilities, 213 are stockholder-owned, and they service nearly three-quarters of all of the electricity customers in the country. About 2,000 public utilities, run by state and local government agencies, supply about 15 percent of customers.

The balance of America's electric power, about 12 percent, is supplied by over 3,000 electric cooperatives, independent power companies, or customer-owned distributed energy facilities.

Ideally, a power plant is located centrally to the community it serves. Sometimes this means that a plant is located close to a city and its surrounding suburbs. In other cases it means that it's located at the midpoint of a triangle formed by three cities. In practice, however, the location of a power plant is influenced by many other factors beside proximity to its end users, such as environmental considerations or whether there is a nearby water source for cooling. Often a determining influence in where to site a power plant is the ready availability of a fuel source. Coal-fired plants, for instance, are frequently seated near coal mines. This reduces the cost of transporting the coal great distances— saving truck or train fuel and the accompanying carbon emissions in the process—but it also creates other problems.

The average fossil fuel–fired power plant has a delivered electricity efficiency of about 33 percent. Why such a low figure? Part of the reason is the long distances that power has to travel to its end users over the grid system.

The grid actually consists of two separate infrastructures, the transmission system and the distribution system. The transmission system carries the electricity from power plants, along a network of over 157,000 miles of high-voltage cables, to lower-voltage distribution systems that take the electricity from transmission lines and deliver it to your home or business. The use of high voltage helps to minimize electrical losses along the miles the power must travel before it reaches its destination at the distribution system, but it's impractical for the distribution lines themselves. At substations, which is where the two grid infrastructures interface, the electrical voltages are stepped down through transformers, from 765 kilovolts (kV) and 138 kV to 120 or 240 volts, for general use.

Think of the high-voltage transmission system as the trunk of a great electric tree. The places where the limbs fork off are the substations, which in turn fork and divide into the branches, twigs, and leaves that are the homes, businesses, and industries receiving the electricity. As nutrients course up a tree's trunk from the sources in its soil, traveling to each hungry leaf, so electricity travels from the "soil" of power plants, along transmission lines, through transformers at the substations, and along tens of thousands of feeder lines into individual homes, businesses, and industries.

The benefit of this delivery system is that power from this grid can come to the individual buildings from a variety of sources—coal or natural gas or petroleum, nuclear, and renewable resources too—depending on what fuel is most readily available to the power plant operator at the least cost. This mix of sources is referred to as system power. While there are ways to tie an individual user financially, and contractually, to a specific power source, as the merchants on U Street are tied to wind power, it is still the case that all electricity drawn from the grid comes from system power.

Another function of this system, in the past, was to regulate the price of electricity. But at the same time that it kept prices level, it kept the customer from being able to choose power providers and power sources, and it kept the price of electricity from being competitive.

In 1992 the Energy Policy Act was passed. Distributors were no longer limited to buying from the nearest power plant in their own geographical area. The new regulations made it possible for generating companies to sell their product for the best price they could get and for distributors to buy at the lowest cost possible. In theory, these new rules were supposed to lower the price of electricity to the consumer as the competition worked to lower the price to the distributor and the savings were passed along.

But what happened was that, as the power supplier in South Carolina offered a better deal to the distributor in Wisconsin, or the distributor in Minnesota got a price break from the supplier in Louisiana, electricity had to travel even longer distances along the high-voltage network, with even greater losses along the way.

At the same time, demand for power was, of course, increasing—25 percent since 1990—but construction of new infrastructure was declining by 30 percent.

Now imagine that over the last twenty-five or thirty years, hundreds of new limbs have been grafted onto the grand old electric tree. Hundreds of new power plants, hundreds of new substations, hundreds of thousands of electric customers located in suburbs ever farther from the sources of their power finding hundreds of thousands of new ways to use electricity as they acquired home computers and fax machines and cell phones and microwave ovens . . .

As the bulk power market grew, electricity suppliers that had once functioned more as colleagues than as rivals began to compete with each other for use of a finite number of transmission and distribution lines even as the number of transactions grew. For example, the annual number of electric transactions on the Tennessee Valley Authority's (TVA) system was less than 20,000 in 1996; today they number over 250,000 annually. This is a volume that the system was not designed—and is not equipped—to handle. The old electric tree is severely stressed.

This stress is a matter of simple physics: The old lines simply can't handle the modern load they are expected to carry.

The distribution system is divided, both physically and administratively, into three "interconnects": the Eastern, which covers the eastern two-thirds of the United States and Canada; the Western, which covers most of the rest of both countries; and the Electric Reliability Council of Texas, which covers most of Texas. All of the generators within each interconnect must remain tightly synchronized within very narrow limits to the same cycle of 60 hertz. A hertz refers to the frequency of an event per second—one hertz means that an event happens once a second, sixty hertz means that the event occurs sixty times in one second, and so on. Even a small frequency change can cause instability in the grid. A generator that drops even 2 hertz below the 60 hertz standard can build up enough heat to destroy itself, causing the link to sag or break, creating voltage fluctuations, tripping circuit breakers at substations, sending overflows into neighboring lines and causing them to work above their capacity. The neighboring lines will in turn build up heat that presages chain reaction failures: dangerous grid congestion, bottlenecks, what we know as brownouts and blackouts.

On the plus side, the grid system is redundant, meaning that there is not just a single line on which electricity can be sent to any customer. There is usually an alternate route around a line that's overloaded and not functioning properly.

The downside is that where to send the electricity—over which line—is a snap decision that has to be made by operators who are often in competition with each other for use of the same line. And they are handling a product that has the shortest shelf life of any in the marketplace.

Electricity is critical to daily life and economic stability, and, as we talked about in the last chapter, there is no practical way to store the stuff. It is the quintessentially fresh product, made for you just minutes before you use it. Power suppliers must decide where to buy it and how much of it to buy based on a set of complicated conditions, including market prices of fuels and anticipated use based on variables such as local weather forecasts. Then they have to coordinate where to send it over the grid, often within seconds, to users who are, just as often, paying a fixed rate per kilowatt-hour.

How does congestion on the grid impact the price of power? On a hot day, you turn on your air conditioner expecting that you will pay marginally more on your electric bill at the end of the month, reflecting the extra units of electricity that you've used.

But all your neighbors are turning on their air conditioners too.

One extreme example of how air conditioning on a hot day can impact an electricity supplier took place in the Midwest in June 1998. The typical cost of buying electricity is about 2.5 cents per kilowatt-hour, and it is delivered to the customer at a cost of between 7 and 10 cents. But on that one hot day in 1998, when everyone wanted electricity to turn on his or her air conditioner at the same time, lines to transport electricity were at a premium on the congested grid, and the typical cost to the distributor rose to $7.50 per kilowatt-hour.

Let's put that $7.50 into terms we can all immediately grasp. If you or I walked in to the corner store to purchase a pack of breath mints that normally cost $3 a container and found that the store wanted $900 for it, we'd hurry to a different store. But the Midwest utility had no luxury of shopping around. It had to either pay the going rate or leave its customers without air conditioning on the hottest day of the year.

The total cost to the utility as a result of the price spike was estimated to be around $500 million.

Finally, we aren't finished with all that we ask this already stressed system to do for us. As car manufacturers move to respond to the public's call for cleaner car technologies, plug-in hybrid vehicles are set to hit the market in a matter of years. Coming home from work in the evening and plugging your electric car into a wall socket to recharge overnight could be exceptionally convenient—and it could also double, or even triple, your home's electrical use.

So, how do we fix the Rube Goldberg contraption that was once the world's standard for electrical transmission?

One way, which has been supported by the Department of Energy (DOE), is to expand the transmission capacity of the existing grid. This would involve building new transmissions lines—and easing the environmental regulations that now limit their construction. But the costs to do this are estimated to be in the tens of billions of dollars, and the DOE proposes to pay for it by increasing utility rates. "The people who benefit from the system have to be part of the solution here," former Energy Secretary Spencer Abraham has said. "That means the ratepayers are going to have to contribute."

But mindlessly expanding the existing grid without bringing a fresh and long-term view to our energy challenge is like continuing to blithely tread water in rapidly rising floodwaters. Upgrading the existing transmission system would take many years, and—rather like a congested highway when a new lane is being added, when the extra lane is finally opened traffic increases almost immediately, so there's no perceptible relief in the flow—it's likely that expanding the existing grid would only just keep up with increasing demand. We would almost certainly be facing another immediate need to upgrade and, in any case, we would still be laboring with the same problem of having to send electricity over long distances in order to get it to the customer.

So, how do we stay connected to the grid without tangling ourselves ever more tightly in its overworked transmission cables, distribution cables, and feeder lines?

All right, Mr. Mayor, here are two solutions you can take to heart.

Smart meters are a smart first step.

Smart meters are, simply, just like your regular electric meter, but with a brain. Rather than merely recording the amount of power electric customers use, smart meters are equipped with a readout that lets customers know the real-time price of the electricity they are using or are about to use.

How does that help to provide relief to the overtaxed grid? Supply and demand.

Electricity is more expensive during peak hours, when there is the highest demand for it and distributors are jockeying for use of limited delivery lines and cables. When electricity is less expensive, generally overnight, it's because demand is down and more lines are open to transport power. On average, in fact, overnight consumption is only 25 percent of what it is during the daylight hours. Smart meters allow electric customers to choose the price they are willing to pay for electricity. They also tip customers off, by way of a higher or lower price, to the real-time activity on the grid and allow them to actively participate in easing any congestion by opting to perform nonessential tasks that require electric energy at times when its price is low.

As we just noted, electricity is generally less expensive during off-peak hours—overnights. Even without a smart meter, an important part of your contribution to energy conservation can be to wait until just before you go to bed for the night to run the dishwasher or start a cycle in the washing machine.

With smart meters, however, consumers have a real tool to use that connects their home—and their *awareness*—directly to the activity going on at any given moment in the second-by-second paced energy market. Consumers get to be players—making snap decisions that can lower their monthly electric bill, conserve energy, and relieve grid congestion in an important way—and all of it from the comfort of their own homes.

The second solution we want to talk about regarding grid congestion is one that can be an inherent part of our inevitable transition to renewable power sources. And the more wisely we think through and implement this transition, the more far-reaching it becomes. You know that old saying, "Think globally, act locally?" Well, it is especially relevant as it applies to how we design our future electric delivery infrastructures. The more local solutions we can implement to relieve stress on the grid, the better our national grids will function.

And how will that work?

We asked you to imagine yourself mayor of your town in the hopes that by inviting you to view the world's energy challenge within a very localized framework, you would be inspired to see the untapped opportunities for energy independence that really are all around us. Perhaps in your community there are one or more of the 77,000 dams in the United States that are not yet retrofitted for hydroelectric generation. Maybe there's a hot springs nearby that is ripe for service as a source of geothermal power. Perhaps there's a sun-baked stretch of land just outside of town, unsuited to agricultural use but just right for a solar chimney. These are all potential sites for a locally-generated supply of energy. And when energy is manufactured within or near the community that will use it, the less distance it has to travel to its destination—and the less distance it has to travel, the fewer transactions are required across the grid. Fewer transactions means less stress on the old electric tree.

Energy experts are in fact urging us, as we build the new energy delivery infrastructure we need to meet our modern energy demand, to rethink the structure of our vast, contemporary central grids. They are telling us that smaller power sources located closer to the consumers could relieve the bottlenecks associated with our current method of delivering power over long distances via high-voltage lines. These small sources of power could reduce transmission losses and yield enhanced power quality. The "problem" of remote locations sometimes necessary to take advantage of natural renewable power assets could actually work in our favor by forcing us to wean ourselves from old ways of thinking about how our grid is structured, toward the development of an infrastructure as evolved as our new energy sources.

There will likely always be some form of centralized grid systems, because there will always be remote locations, such as the Mojave, that are naturally suited to producing energy in large, utility-scale quantities. But renewable energy technologies lend themselves to use at scales that take advantage of local and regional geographical assets and weather conditions. They allow a mayor to envision his or her community as becoming more and more self-sustaining with each solar roof installation or each turbine turning in the breeze. Each swimming pool equipped with a passive solar water heating system requires just a little bit less coal to be trucked to the power plant, and each shovelful of coal spared at the power plant saves the grid from that kilowatt or two that might trip a circuit breaker and start the chain reaction blackout.

If you're like the majority of people to whom we've posed this "mayor for a day" scenario—and we suspect that you are because, in the very act of picking up this book, you've shown your desire to find answers to, and to be a part of, solving the world's energy dilemma—by now you've got a lot of ideas percolating about the sort of best-case solutions that can be implemented, and with all due expediency, in your home community. And, if by chance, you really are a mayor, or a city councilperson, or a member of, say, your local school board, then you're already in a position to propose the solutions you've thought of and foster the cooperation among your fellow elected officials and constituents that will drive the momentum to see them become realities.

But what if you've never been involved in public life in your life? We're suggesting that now is the time to step out. This doesn't necessarily mean that you have to run for elective office. It does mean that you can seek out the opportunities you have to make your own home or business more self-sustaining—whether that's taking advantage of an incentive program to install a solar heating system on your home or business building or simply waiting until nighttime to start your washing machine.

You can also take your idea or ideas to the elected officials you already have by asking for a place on the agenda at the next public meeting.

You can actively support the election of politicians who embrace the best practices we outline in chapter 10, by donating time or funds to campaigns. And you can join an organization that monitors the progress of energy legislation and that will alert you—often conveniently via e-mail—regarding bills coming up for debate or vote so you can add your voice to the public's insistence that no more time is wasted in acting on practical and timely solutions to the coming energy crisis.

We're hard-pressed for ways to state this more plainly or strongly: It's individuals keeping up the pressure on their leaders and showing them the way that is going to move energy solutions beyond talk and forward as quickly and innovatively as we need them to move. After all the creative exercise of playing mayor for a day, our energy problem isn't a game, of course, at all. Unlike in a SimCity society, we can't just erase a lethally polluted town from our game disc and build a new one.

6

FUELISH CHOICES

In 2007 Brian Schweitzer, the governor of Montana, came to New York on business. One of his meetings on that trip was with Michael, ostensibly to discuss using coal to produce liquid fuel—a technical discussion about a capital-intensive project—but almost before they had a chance to greet each other, the governor had placed a handful of small seeds down on Michael's desk.

"They looked like mustard seeds to me," Michael told me. But Governor Schweitzer set him straight.

These were camelina seeds—a plant that is, in fact, related to mustard, but it is not a food source. Camelina was cultivated widely in Europe during the Bronze and Iron ages, prized for the oil it released when it was pressed, and the governor was promoting this ancient crop now to Montana's farmers. This plant, the governor explained, could thrive in dry conditions, so it didn't need a lot of irrigation, which was good if water got as scarce as some people said it was going to get, and camelina had some big advantages as a rotation crop too—crop rotation being the standard agricultural practice of planting dissimilar crops in the same acreage from year to year in order to avoid the soil depletion and build-up of pests and pathogens that often happens when one species is continually planted in the same space. The governor's goal was to get 50,000 acres of high-yield strains planted in his state.

Michael naturally wanted to know why.

The governor smiled. "Biodiesel."

Now we come to the third and final piece of the energy puzzle that must be put into place to secure our energy future: How are we going to fuel our cars in the coming renewable age? The answer to this question is, as you'll see, rather complex, but we'll start the discussion with an obvious declaration: We don't need gasoline to drive our cars.

Plain as that.

We have biofuels. We have ethanol, which is made by fermenting the sugars contained in plant matter, as an alternative to gasoline. And we have biodiesel, which is made from oils that can be extracted from plants, as an alternative to diesel fuel oil.

With oil prices hitting $100 a barrel and world oil supplies moving fast toward empty, it may well strike us breathless with relief that we have alternative ways to fuel the cars, trucks, and minivans that shuttle us around the world. But the plain truth is that we don't need oil to stay on the move and *we never did*.

In 1898 Rudolph Diesel demonstrated his invention, the compression engine, at the World's Exhibition in Paris—and peanut oil was the fuel that he used. In 1908 Henry Ford introduced the Model T—and it ran on ethanol. These automotive pioneers used plant-derived fuels not because gasoline wasn't available in their time, but because they presciently believed in building machines that ran on renewable fuels rather than the resource-consuming steam engines of the day that, as the industrial age progressed, grew hungrier and hungrier for coal, filling city skies with smoke and noxious fumes. Ford, in fact, was so convinced that energy produced by renewable, plant-based fuels was the future that he built an ethanol-processing plant in the American Midwest and partnered with Standard Oil to sell biofuel in the oil company's emerging network of fueling stations. Through the 1920s, ethanol accounted for a full quarter of Standard Oil's fuel sales.

So, what happened?

If we owe car engineering excellence to the Germans (and we do—to inventors like Gottlieb Daimler and Karl Benz), and if we owe a nod to the French for taking on the risk of manufacturing a branded line of that new invention, the car, and making it available to the public (and we do—Peugeot was the first successful commercial automobile manufacturer, and the company is still going strong today), we owe the creation of a "car culture" to the United States. Henry Ford's assembly-line approach to carmaking, along with the Model T's affordable price tag, made the luxury of motoring democratic. By 1927, 15 million Model Ts were on the roads, and most of them had been built from the chassis up, ready to roll out of Ford's automated factory, in just ninety-three minutes.

But, at the same time that the world was discovering the joys of the open road, something else—something else very uniquely American—was going on: Prohibition.

Prior to the end of World War I, the ethanol industry in America had been a booming business, with nearly 50 million gallons produced every year. Ethanol was used not only as an automotive fuel but as a material in the making of a wide variety of products that were necessary to the war effort. Then, in 1919, the Prohibition era began. Ethanol is, as we've said, a product that is made from the fermented sugars of plants, much in the same way that wine, beer, and whiskey are produced, resulting in alcohol. During Prohibition, it was illegal to manufacture alcohol; ethanol, because of its alcohol content, was considered liquor. The only way that it could be sold within the law was if it was mixed with some poisonous substance that would make it unfit to drink. The poisonous substance that it was mixed with was petroleum.

The era's oil barons quickly took advantage of the disfavor in which alcohol was held to discredit ethanol and consolidate the place of gasoline in the new car culture. All it took was a few minor adjustments to the car's engine design, and abetted by the fact that oil at the time was cheap and plentiful—presto! By 1933, when Prohibition was repealed, ethanol was thought of as a gasoline extender, or maybe an octane enhancer that boosted a car's performance, but it was no longer thought of as a fuel.

Standing, as we are, at the dawn of new energy age, faced with rapidly depleting oil supplies, we have no choice but to think, once again, of ethanol as a fuel. And to make the Diesel/Ford dream of a biofueled future ours at last.

BIODIESEL

What is it that prevents us from pulling up to any fueling station of our choice and filling up our tanks on environmentally-friendly biofuels? Both ethanol and biodiesel are, after all, made using the same basic science that was known long before the turn of the twentieth century. And because awareness of the coming energy crisis has lately focused our attention on alternative technologies, considerable advancements have been made in the field that ought to make for an even smoother transition to green fuels. Shouldn't we—*don't* we—have the scientific wherewithal to, at the very least, *begin* to integrate biofuels in some comprehensive way into the plan for our energy future?

Yes. We do.

And a slow, steady integration is pretty much what's happening in the case of biodiesel.

Diesel engines were tapped for use in motor vehicles from Rudolph Diesel's first peanut oil–fueled design because they could be made to run on a petroleum product that was, in essence, a by-product in the manufacture of gasoline. The mechanics of Diesel's engine allowed it to run on what can be described simply as a lower grade of fuel. By promoting the manufacture of cars with diesel engines—vehicles that could be powered with what came to be known as diesel fuel oil—the oil industry created a market for this secondary product, and the overall value of the gasoline refining process was increased by giving oil distributors another product to sell.

*Bio*diesel is made from the oils that can be pressed from plants like peanuts, soybeans, palm, or camelina, or from other fats, including animal fat and recycled restaurant french fry grease! As fortune and the science behind what makes biodiesel work would have it, biodiesel is free of two of the major issues that have so far prevented a wide and rapid adoption of ethanol use. Those issues are an infrastructure that can handle the delivery of the fuel to its market and vehicles that are equipped with engines that can run readily on the renewable fuel.

Nothing has to be done to a diesel engine car or truck in order for it to be able to run on biodiesel. In fact, a great deal of diesel fuel on the market today is a blend of standard diesel fuel oil and biodiesel, often about 5 percent. There is no difference in how a vehicle performs when these blends are used. For blends with up to even 20 percent biodiesel content, most manufacturer warranties on the vehicles are still valid. Indeed, new diesel engine cars, whose mechanics haven't yet been impacted by a steady diet of old-fashioned diesel fuel oil, can be run on 100 percent renewable fuel from day 1 without a hitch in their performance.

Now, if you're going out tomorrow to buy a new diesel engine car, what might keep you from running it entirely on a soybean or camelina seed product is a lack of access to a reliable source of 100 percent biodiesel. That is slowly starting to change. Biodiesel, unlike ethanol, can be transported in the same pipelines that are already dedicated to fossil fuel oil without concern that the new fuel's chemical makeup will corrupt metal pipe material. Because it can be more readily moved around within the existing fuel infrastructure, biodiesel is the fastest-growing sector of the renewable fuel market, with a record production of 250 million barrels in 2006 in the United States alone, and plans to move 3 billion gallons of the stuff through the existing infrastructure, to consumers, in the near future.

Because biodiesel is by its nature more chemically and mechanically compatible with the fuel that it is replacing, in both distribution pipelines and in vehicle engines, it's already made a comfortable—if still tiny—niche for itself in the fuel market.

Ethanol is a more complicated story.

ETHANOL

If, in our mayor-for-a-day game, you have been sitting in as the chief executive officer of a city in the American Midwest—a city, say, in Iowa or Nebraska that's surrounded by cornfields—you are going to have your own understanding of what ethanol is. It's likely you'll know someone, or at least know *of* someone, who's involved in the corn-based ethanol business. In fact, nearly 25 percent of America's annual corn crop currently goes to making ethanol.

If, however, you live in Teesside, in England, near to where the country's first wheat-based ethanol plant is now under construction, or if, in our game, you're sitting in for the mayor of a town in Brazil where sugar-based ethanol is king, you might have an entirely different take on the biofuel. There isn't a great deal of room for debate that someday— and the sooner the better—we're going to be driving vehicles that run on biofuels. The conundrum is in agreeing on what the best feedstocks are. Ethanol can be made from an extensive variety of plants, not just corn or wheat or sugar. What are the most energy efficient and cost-effective raw plant materials from which to manufacture our new fuel?

Our perceptions about what plants make the best ethanol are formed in many ways by what crops have historically been abundant in our own geographic areas and by the idea of finding new uses, and new revenue streams, for products—in this case, plants—with which we are already familiar. This is not always the best approach. Now, we still want you to "think locally," as a mayor has to think, while we talk about the best options for energy crops, but we want you also to continue to think out of the box about your area's particular ethanol opportunities. Especially in America, where corn-based ethanol dominates what is the largest ethanol-producing nation in the world, we need to speak very frankly about ethanol feedstocks in order to make wise decisions about our biofuel future. Ethanol technology has progressed, as we'll see, beyond the comfort zone some of us have already established around this new fuel; that is, we've gone beyond corn.

In order to understand how far biofuel technology has come—and how far it can, with appropriate investment in research and development, still *go*—let's think for a minute about the advances that have been made, just in most of our lifetimes, in a wholly different arena: how we listen to music.

When Michael was a teenager, the two pieces of electronic equipment that were essential to a young adolescent were a transistor radio and a portable record player. The transistor allowed you to listen to your local pop radio station, hear the latest recordings by your favorite artists, and decide which of the hits you needed to own on a 45 rpm vinyl disc. By the time he was in high school and then college, component stereo systems featuring turntables and monster speakers were the order of the day, the better to crank up albums by Pink Floyd and CSN&Y—a standard deviated from, and only briefly, by a flirtation with 8-track systems.

By the 1980s, music had become portable—boom boxes and Walkmans. How liberating it was to go to the gym, plug in earphones and pop in a cassette, and work out to your own private programming—my choice was Paul Simon's *Graceland*—rather than to the Top Ten rotation that was the inescapable background soundtrack at health clubs in those days! Those old Walkmans seemed so small and weightless—so revolutionary!

But Walkmans were a small advancement compared to "the next big thing" that awaited audiophiles: the compact disc. Walkmans were gadgets that allowed a listener to transport music, but the music still came on the same old cassette tapes or on the same old vinyl discs. CDs were a whole new format. From the time of the introduction of the CD, new audio technology progressed so quickly that Michael and I never really got around to replacing all of our old 33-1/3 LPs with compact discs before iPods and the ease of downloading a favorite artist's work on iTunes overshadowed the last iteration.

Finally, who among us who started out a lifelong love of popular music with an ear pressed to the static strains of a transistor radio, hoping the DJ would spin our best-loved song again, and soon, will really ever get over feeling at least a smidgen of awe at the four-ounce gadget that calls itself the Shuffle? Two hundred and fifty crystal-clear songs on a bit of plastic and computer circuitry about the size of a poker chip!

We've come a long way, baby, and in a really, really short amount of time.

This is the same steep and speedy trajectory we're headed on when it comes to biofuels. The velocity of the trip has got to be powerful because the health of the planet and of our economies demands it. And the transition *can* be quick because we've proven that, when it comes to embracing ever more efficient and liberating technologies, we're no slouches. The only real thing that can stand in our way is if governments fail to adopt policies that empower the advance of the next generation of cars and the ways that we fuel them.

But government policy is a discussion we'll get to later.

Right now, let's focus on our ethanol options.

All Ethanol is not Created Equal

When many people think of ethanol, they think of the alcohol that results from fermenting the sugars in corn, or possibly in wheat, as these are the feedstocks that get the lion's share of press. Certainly corn- and wheat-derived ethanols make adequate fuels to power our cars. But are they the best feedstocks for ethanol? Are they—or can they become— cost effective? Do they make for as *clean* a fuel as ethanol can be?

First things first: Biofuels are not "clean." That is, all biofuels release carbon dioxide (CO_2) when they're burned. That's natural, as both ethanol and biodiesel are made from plants that are living things, and living things are all composed of carbon. But the quantity of carbon that is released when a biofuel is burned is minimal when compared with the quantity released by a petroleum product. And the minimal quantity of carbon released by burning a plant-based fuel is then reabsorbed efficiently back into the earth by the new crop of corn or wheat or sugar or soybeans or any other plant that is sown to provide the raw material for the next batch of ethanol. The cycle of carbon released through the production of mechanical energy to run a car and the carbon reabsorbed by the cultivation of plants that will ultimately produce another round of mechanical energy results in a stable balance of carbon in the atmosphere. This state of balance—of producing only the same amount of carbon as the earth is able to reabsorb—is what will allow us to drive our cars sustainably when we start to power them with biofuels.

Some plants used for the production of ethanol, however, are responsible for less emissions of carbon and other pollutants than, say, corn is, and not necessarily because the *ethanol* made from them burns "cleaner." It's because some plants require less energy to be expended as they are grown, harvested, and processed into fuel than others.

Remember the artificial Christmas trees we talked about in chapter 3? An analysis of the fossil fuel energy used to make an artificial Christmas tree has got to include not only the oil used to make its plastic parts. It must also take into account the coal used to supply electricity to the factory where the tree was made and the diesel fuel used to power the ship that transported it from its factory to its destination. Unless the fossil fuels used throughout the whole process of producing and delivering the artificial tree are considered, the analysis is meaningless as a measure of the impact the tree has on our energy environment.

So it is that, in determining how appropriate any crop is for use as a biofuel feedstock, one has to consider not just the merits of the final product but a whole range of variables particular to each specific crop, including how its cultivation impacts the soil in which it is grown, how much energy must be expended in the process of turning it into a fuel, and how cost effective it is for the manufacturer to produce and, ultimately, for the consumer to buy the final product. Only by taking into account a whole host of factors like these can we arrive at a meaningful understanding of how practical any particular plant will be as a feedstock.

Let's start with corn. In terms of the music analogy we used to illustrate the progress of technology, we might say that using corn to make biofuel is like listening to Sheena Easton on your old Walkman. The Walkman technology is adequate and serviceable, and it's liberating in the sense that such a personal, portable sound system has never before been available to you. Depending on how inspirational "Strut" was to you as a workout tune, you were probably thrilled to have a Walkman to tote around.

Just as the Walkman was an altogether necessary precursor to today's transportable music gadgets, corn was an altogether necessary, and reasonable, choice of crop with which to begin the biofuel evolution. Corn is high in sugar, and it produces a fuel that's perfectly adequate and serviceable in our cars. Twenty years from now people in the biofuel business will likely be saying that corn-based ethanol jump-started a renewed awareness in the utility of alternative fuels. People were making corn ethanol. People were using corn ethanol. People were investing in corn ethanol. When people start to put money into a concept, others start to take it seriously and the effect snowballs. When we are all driving around in ethanol-fueled cars, it will be corn farmers and corn-ethanol producers we'll have to thank for starting us on the way down this good, renewable roadway.

But the ethanol we'll be using then probably won't be made from corn.

All ethanol is not created equal. A rose by another name might still smell as sweet, but ethanol is *not* ethanol is *not* ethanol, and corn as a feedstock for ethanol has some serious shortcomings. For starters, it's a land- and nutrient-intensive crop, the cultivation of which requires the use of pesticides and fertilizers—pesticides and fertilizers that are made from petroleum. Fossil fuels are also needed to run the farm machinery that harvests the crop and transports it to the processing plant, and a fossil fuel, often natural gas, is used in the conversion process from corn to fuel. Though making corn-based ethanol involves a technology much more mature than hydrogen fuel cells, corn ethanol has something critical in common with fuel cells: It doesn't relieve the need for fossil fuel but merely swaps its use in a car's engine for its use at some earlier point in the transportation chain. The aggregate reduction of CO_2 emissions that results from burning a gallon of corn ethanol versus a gallon of old-fashioned gasoline is negligible. Much corn ethanol today is used as an oxygenating agent in standard gasoline, particularly in areas where methyl tertiary-butyl ether (MTBE)—a highly toxic petroleum-derived chemical—has been banned for this same purpose. Ethanol is the only known substitute. When the amount of oxygenating corn ethanol in the petroleum blend is 10 percent, the CO_2 reduction is just 2 percent. When the ethanol product is E-85—that is, a blend of 85 percent ethanol and 15 percent gasoline, commonly referred to as *gasohol*—the CO_2 reduction for the corn-based biofuel is just 17 percent. We can do better.

We can do better on price too. In order for any ethanol to be ultimately competitive with gasoline, it has to come close to gasoline's price. The cost to produce a gallon of gas averages about 90 cents. Because corn is such a land- and resource-intensive crop, it's unlikely that the cost per gallon production cost for corn ethanol can drop much below the current figure of $2.00.

This base cost of $2.00 per gallon, moreover, takes into consideration the by-products that can come from "dry mill" corn-ethanol manufacture, principally corn gluten mush for cattle feed. By-products lower the per-gallon production cost of a fuel by providing a fuel manufacturer with another valuable product to sell from the same processing cycle, so let's look into them just a bit more deeply. There are two ways of processing corn for ethanol—the "dry mill" method we just mentioned, which produces ethanol and animal feed, and the "wet mill" method, from which more byproducts can be produced than from the dry. One of the byproducts of the wet method is corn gluten fertilizer, a valuable product now in Ontario, Canada, where chemical fertilizers have been outlawed. The downside to the wet mill process, and the reason why wet process refineries are more rare, is that they cost more to build. But byproducts are as important a part in the manufacture of biofuels as they are in the manufacture of gasoline. Just as the by-product diesel fuel oil lowers the cost per gallon to refine gas, and as the residual oils that can come from the manufacture of biodiesel can be used to make pharmaceuticals, vitamin supplements, and cosmetics that bring down the cost of that green fuel, maximizing the use of by-products that result from ethanol production will go a long way toward making it cost competitive. Corn mush just doesn't go far enough.

So, let's think creatively: What is a more naturally economical crop to use to make the ethanol fuel we need for our cars? A crop that requires less acreage to turn out the same amount of energy, and so makes better use of agricultural lands and resources? A crop that has a lot of by-product potential? An energy crop that costs less to grow and, at the same time, increases profits for the energy crop farmer?

Let's think about *sugar*.

In order to do that, we have to go to Brazil.

POWERED SUGAR

Brazil is the world's largest producer and exporter of sugar and, behind the United States, the world's second largest producer of ethanol. Though the volume of ethanol produced in both countries is about the same, Brazil uses less than half the land that the United States uses to grow its ethanol crop because, bushel for bushel, sugarcane simply yields more energy than corn. And, in contrast to the United States, where ethanol use is but a trickle of about 3 percent, ethanol accounts for over 40 percent of the total motor fuel consumed in Brazil.

How is it that Brazil came to be the world's largest *user* of ethanol fuel?

A fair analysis has to disclose that the Brazilians are working with some advantages that most of the rest of the world can't duplicate, at least as they apply to sugar. Brazil's equatorial climate is a natural fit, of course, for sugarcane cultivation; planting sugar in rain-fed fields, for instance, eases considerably the costs to grow what is a water-intensive crop. Labor, as well, is cheap in Brazil, so the costs to harvest sugar, which is in fact a *labor*-intensive crop, are reduced. And about that factoid that ethanol makes up 40 percent of the total motor fuel consumed in Brazil versus America's measly 3 percent? It has to be told that Brazil must furnish significantly fewer vehicles with fuel than does the United States: approximately 23 million vehicles as opposed the U.S. fleet of over 204 million.

But in pointing out Brazil's advantages, we don't want to underplay its enormous accomplishment either. Brazil has produced the first renewable fuel to be cost competitive with its fossil fuel counterpart. Sugar-based ethanol has a per-gallon production cost of around 75 cents. At the pump, Brazilians can pay either what translates at the exchange rate that prevails as of this writing (U$1=R$2.15) to $4.20 per gallon for gas, or they can pay $2.70 a gallon for ethanol. Even with the energy efficiency of ethanol versus gasoline, which averages about 70 percent, ethanol is still a bargain for Brazilian motorists.

Wait a minute. Back up. Brazilians can drive into any fueling station and have a choice to fill up with ethanol?

They can, indeed. Brazilians have the tools—the gasoline pumps and the sorts of cars—that allow them to make more informed choices about how they will fuel their vehicles. Brazilians don't shop for gasoline or ethanol so much as they shop for value in Btus. The price of each commodity, consequently, moves with the other. That is, for example, if the price of sugar skyrockets, people will buy gasoline and, if the price of crude goes up, they will buy ethanol.

To discover how Brazilians have managed to produce a price-competitive product and integrate it efficiently into their fueling infra-structure, we have to look beyond natural advantages like a climate suited hand-in-glove for a high-yield energy crop like sugar. We have to dig a little deeper for the lessons that the rest of the world can take away from Brazil's ethanol success story. It wasn't just the sugar but the intelligent and comprehensive plan for how to use it that made for Brazil's biofuel home run.

The first smart thing the Brazilians did was to build their ethanol refineries right on the sugar plantations. This close proximity of field to refinery eliminated the cost—and the use of fossil fuel—to transport the crop. The natural question to follow from this statement is, But in the United States, aren't biodiesel distilleries planted in soybean fields, or ethanol refineries in corn fields? Yes, they are. But the same logic of logistics hasn't worked out as well for soybeans or corn for several critical reasons we'll discuss in the next few pages.

The second smart thing the Brazilians did was that they generated the electricity to run their sugar refineries with *bagasse*. The sugars in a plant's "fruit," from which ethanol is produced in most technologies, account for a little less than a third of the energy stored in the whole plant. Two-thirds is stored in the stems and leaves and the bagasse—the fibrous material that's left over after a plant is pressed. By burning sugarcane bagasse to provide the electricity to run the machinery at their sugar distilleries, ethanol plants became energy self-sufficient—and created at the same time a by-product that can do so much to lower the cost of the primary product, ethanol, to the consumer. In this case, the by-product is electricity itself. The Brazilian sugar refineries are able to sell the excess electricity they generate by burning bagasse to their national grid. It's estimated that currently the refineries generate 600 megawatts from bagasse for their own use and 100 megawatts to sell to the grid.

Finally, Brazilians began to aggressively research and develop other potential by-products, beyond bagasse-generated electricity, that would slice the costs of their ethanol production even more. So far these by-products include the promise of the bioplastic we spoke about in our introduction—plastic that will allow you to preserve a head of lettuce in your refrigerator, or slake your thirst with a long, tall drink of bottled water, or sanitarily pick up your dog's poop and then will dissolve, in a brief span of time in a landfill, right back into sugar. Promising experiments being conducted even as you are reading these words aim to turn oils from sugar into jet fuel. In short, what is underway is the concentrated search for ways to extract from sugar the building block molecules now derived from chemical processes, to produce a wide range of the products we use every day—from plastics to cosmetics to jet fuel—and to make them better by making them green.

None of this happened in a vacuum. It took a deep commitment to a comprehensive plan to change the way Brazilians drove, and it took money. But in 1975, when the country's National Alcohol Program (PROALCOOL) was launched to support its ethanol initiative, the oil embargo by the Organization of Petroleum Exporting Countries (OPEC) had cast fears that oil dependency could endanger national security and fostered the impetus to act. The oil embargo also coincided, rather fortuitously as it turns out, with a sharp fall in sugar prices in 1974, and the government had to look for a way to guarantee profitability for one of the country's primary industries. The inspired choice was to go into the production of sugar-based ethanol.

The first target for PROALCOOL was to increase the number of ethanol distilleries by offering low-interest loans for their construction. Then the government charged the state-owned oil company, PETROBRAS, to make the investment in infrastructure that would make possible the distribution of ethanol to the consumer; this was managed through a subsidy plan. In a little less than four years, ethanol production was nearly three times what it had been when the program was launched.

The final step in transitioning Brazil from a gas-fueled country to one that was renewably mobile came in 1979, during the Iranian revolution, when a second oil crisis deepened fears for energy security and—not at all beside the point—increased the price of gas. This step was to start promoting the production of cars designed to run on ethanol. The government signed agreements with major carmakers, including Volkswagen, GM, and Toyota, to produce ethanol-fueled vehicles. It gave tax breaks to taxi drivers who converted their cabs to run on 100 percent ethanol. It mandated that its own fleet be all-ethanol.

Because of this strong government initiative, by 1988, a stunning 90 percent of new car sales in Brazil were of the ethanol-fueled kind.

Brazil's transition to ethanol didn't happen without a few bumps, of course. In the mid-1980s world oil prices dropped sharply, and with them the benefits to the consumer of using ethanol over gasoline. Though Brazilians noted with dismay the change in air quality in their larger cities, like São Paulo, which had cleared up in the ethanol days and then worsened again in proportion to the return to the use of fossil fuel, there wasn't that immediate economic incentive at the pump for people to keep choosing a fuel blended with a high percentage of ethanol.

Then, in 1988, world sugar prices *rose* sharply. To maximize the value of their crops, sugarcane planters diverted harvests from ethanol distilleries to sugar refineries for export as a food product. The result was that Brazil faced a severe ethanol shortage and actually had to import the fuel. In response, drivers stopped filling up with ethanol for a while, and carmakers stopped making ethanol-fueled vehicles.

This experience of sugar competing in two different markets illustrates clearly one of the risks involved in using certain plants as ethanol feedstocks—and one of the problems that has beleaguered corn and wheat ethanol producers in other countries. Sugar, corn, and wheat are not only fuel; they are food.

Brazilians have since aggressively tackled the problem of sugar-as-food—in fact, as they have entered into ethanol export agreements with Germany and Korea, and geared up to increase production accordingly, they've greatly expanded the area devoted to sugarcane plantations. Currently, about 14 million acres are devoted to sugarcane production, and this acreage is about evenly divided between ethanol and sugar production. A study from Iowa State University estimates that there are about 250 million acres of degraded pastureland and about 225 million acres of savannah that could be used for the production of sugar ethanol. This is a total potential of 475 million acres of land for fuel that could convert into the energy equivalent of 205 billion gallons of fuel a year. For a perspective on how big a deal this is, consider that in the entire United States 120 billion gallons of gas are consumed annually.

But in the United States, prime agricultural land suited to corn cultivation is limited. Increasing the demand for corn by using it as an energy crop drives up the price of corn as food—and it's not just the price of an ear of corn or a bag of tortilla chips we're talking about. Corn is a primary feed for livestock; the next time you're wondering why your steak is so expensive, remember that the market for corn as ethanol puts a premium on corn as food, including animal food, and the rancher's extra cost for feed is passed along to you at the grocery store.

We don't want to gloss over, either, the steadfast commitment that was required to make the physical changes to Brazil's infrastructure that now support a widely reliable ethanol supply to drivers. These changes were not easy to accomplish. For a while, back in the 1980s when oil was cheap, Brazil was even a laughingstock on Wall Street for its dogged determination to transform itself into an ethanol-fueled nation, and that couldn't have made getting over the hurdles the country faced any easier. One of the most challenging was putting into place the pipelines that would move ethanol conveniently through the distribution chain.

Right now, for example, in the United States, much gasoline already contains about 10 percent ethanol, and according to at least one industry insider, even the addition of this small fraction to the gasoline supply was "very painful."

The practice of blending 10 percent ethanol into gasoline came about mostly because of mandates arising from air quality problems in large metropolitan areas caused by pollution due to the oxygenating additive

MTBE. Gasoline is shipped around the country via metal pipelines; ethanol, however, because it attracts water, is corrosive to certain metals. For this reason, the pipeline owners are naturally apprehensive about introducing it into their existing lines. So, if the ethanol can't go through the pipelines, how can it get to the "racks" (the central locations where fuel is loaded into trucks for distribution to individual filling stations)? In the transition from the MTBE additive to the ethanol blend, the blending had to be accomplished in the trucks themselves, a painstaking and cumbersome process that drivers came to refer to as "ethan-hell."

But let's face our limited options. We cannot continue to power our cars with fossil fuel. Even without the problems created by greenhouse gas emissions, oil supplies aren't going to last more than another fifty years, at the outside. Mass adoption of electric cars will stress grids that already have a hard time carrying their current loads. As for hydrogen-fueled vehicles, well, even given the need to reorganize and augment existing fuel delivery infrastructures to accommodate the transport of ethanol in dedicated, ethanol-grade pipelines, biofuels still have it all over hydrogen fuel cells.

Hydrogen technology is, as of this writing, still another fifteen to twenty years from maturity, and the fueling stations that would be necessary to make the hydrogen cars practical would require the construction of a completely *new* infrastructure. Biofuel infrastructures, however, are fundamentally in place now. With the right leadership enacting wise policies, the reorganization and augmentation necessary to make the existing systems viable for biofuels can be accomplished in a mere three to five years. Given the state of the health of the planet, we don't have the luxury to wait another fifteen to twenty years for a solution to our fossil fuel transportation problem. Biofuels are our bird in hand.

Before leaving Brazil, we need to acknowledge the enormous patience the country has demonstrated as its use of ethanol has grown and evolved. Remember that its ethanol initiative began over thirty years ago! One of the facets of the transition about which, to our minds, Brazilians have been particularly patient, and prudent, is the conversion of their national fleet to ethanol-capable vehicles. This is a conversion, you'll recall, that started in 1979, was derailed in 1988, and didn't really take on new life once more until 2002, when the rising price of oil and the new awareness of global climate change combined for a resurrection.

Today, many places in the world have already begun, as we've said, a slow transition to ethanol use by requiring that each gallon of petroleum fuel is blended with about 10 percent ethanol for pollution-control purposes. Probably more frequently than we imagine, motorists don't even know they're buying ethanol with their gas, and their cars perform just fine.

But for a car to run on a blend with a higher percentage of ethanol—say, E-85 or, ideally, 100 percent ethanol—the engine has to be modified. The fuel pump pressure has to be increased, the battery has to have a higher capacity, and the ignition system has to be recalibrated, among other adjustments. But these are all minor alterations that add little to the cost of manufacturing a new vehicle. Still, of course, it's unrealistic to expect that everyone can simply run out right away and buy a new car capable of handling gasohol the minute that it becomes available on a widespread basis. The manner in which Brazil handled the mix of ethanol-capable and old-fashioned gas cars on its roads is an ingenious model for the rest of us.

When a driver pulls her car up to a fuel pump in Brazil, that's no ordinary pump she's approaching. That pump has a *dial*. Depending on the ability of her car to run on ethanol, or perhaps on the price of ethanol versus gasoline on that particular day, she adjusts the dial to dispense the appropriate ethanol or the ethanol/gas blend and pays the market price for each percentage of the product that she's chosen. If the driver has a Flex Fuel Vehicle (FFV), she's got even greater latitude to decide what blend of fuel is right for her on any given day.

A FFV is a car that comes equipped with an onboard computer with a special sensor that recognizes the blend of fuel being dispensed into the tank and automatically adjusts the engine's combustion parameters *without any input from the driver.* This revolutionary sensor can be added to the vehicle at minimal cost, so FFVs in Brazil, where they were introduced by Fiat in 2002, have turned out to be affordable, popular family cars. How popular? In 2003, the first year after their introduction, 4 percent of the cars being sold in Brazil were the new Flex Fuel models. But just three years later, the percentage had jumped dramatically, to 70 percent. Today, about 85 percent of cars in Brazil are FFVs. Brazilians have so heartily embraced their renewable fuel freedom that one automaker, VW, is phasing out its sales of gas cars in that country entirely.

Throughout the Brazilian ethanol industry's hard times, however, the government continued to support the country's renewable fuels policies. Brazilian leaders sensed very early on the tangible rewards that could come with pursuing a green fuel policy, and they had the will to follow through on the sound instinct. A criticism often leveled at the manner in which Brazil handled its transition to renewable fuel is that it was managed under a dictatorship, but that is, really, entirely beside the point. If it weren't, other dictatorships could then boast the same accomplishment, and they can't. Today Brazil flourishes as a vibrant democracy. Brazilian cities enjoy clean air once again, thanks to the broad use of renewable fuel. The country enjoys a thriving new sugar-ethanol industry, one that is now branching into exporting and is energized to do the research and development that will expand the efficiency and profitability of distilleries by the growing discovery of new by-products. Citizens enjoy all the new jobs that come with a thriving new industry. And, as of 2007, they get to enjoy the realization of a thirty-year-old national goal: independence from foreign oil. It's estimated that Brazil has so far saved $50 billion because it no longer has to import oil.

All of these pleasures came by way of a concerted, thirty-year, $30 billion national commitment to renewable fuel. But, by the way, all of this continues now without government intervention. The Brazilian sugar-ethanol industry no longer has to rely on government support mechanisms. There are no production subsidies, no indirect costs being picked up by other sectors. The only continuing contribution Brazil's government makes to the sugar-ethanol industry is to require that a minimum percentage of ethanol is blended in all gasoline sold in the country, and it does this less as a stabilizing measure for the sugar industry than out of environmental concerns. These days the Brazilian government's role in the ethanol business is to take satisfaction in the model it's provided to the rest of the world in how to sustain the will to attain a sustainable and energy-independent way of life.

How can the rest of us catch up to Brazil? On many levels Brazil enjoys a unique set of circumstances conducive to the production and use of biofuels that, as we've said, the rest of the world can't duplicate.

First of all, most of the rest of the world doesn't have the excellent growing conditions Brazil has for a high-energy crop like sugar. How do we compensate for that? The best initial step we can take is to frankly recognize that very limitation and to circumvent it by expanding the sort of crops we rely on to produce our biofuels. Depending on a single feedstock such as soybeans to produce biodiesel or corn to produce ethanol has led to great struggles in the nascent biofuel industry.

The second benefit in diversifying the feedstocks we are capable of using to produce biofuels is that we can move away from dependence on food crops as fuel crops. The land that is available for use as sugar ethanol acreage in Brazil is extensive. The land for growing, say, corn or wheat in other parts of the world is more limited. This limitation in the amount of corn or wheat that can be grown has a negative impact on the price of feedstocks that are needed to not only power car engines but to nourish people—the more precious a commodity becomes, the higher the cost of it; the more the raw corn costs, the more it costs to make corn ethanol and the less corn ethanol is able to compete on a cost basis with gasoline.

It's not only the feedstocks, however, that we have to consider in our efforts to build a viable biofuel future. The logistics of transporting the finished biofuel to its end user is a problem that plagues the industry in, for example, the United States, where most corn ethanol refineries are located in the midst of corn country, the heartland of the nation, but where the need for the fuel is centered thousands of miles away, on the nation's coasts. In Brazil, this logistical problem was addressed through the construction of an infrastructure—non-corrosive pipelines and ethanol filling stations—through which the ethanol produced at rural refineries can be easily transported to—and purchased by the consumer in—the urban centers where it is most in demand. The rest of the world has yet to make the commitment to building the infrastructures essential to support a thriving biofuel future.

Even as we contemplate upgrading our approach to biofuel through diversifying feedstocks and committing to build ethanol-friendly infrastructures, we can't become so enamored with the concept of green fuel that we overlook the impact increasing our biofuel production and consumption could potentially have in other areas. A study released in January 2008 by Princeton University's Woodrow Wilson School of Public and International Affairs warns us that changing the ways in which we use land to accommodate increased cultivation of fuel crops could have a negative effect on the environment. We have to be thoughtful and remain alert to the many ways our energy future will shape life on Earth. It's considerations such as these that seem to point a sharpened arrow at a type of biofuel we haven't yet talked about: Cellulosic Ethanol.

CELLULOSIC ETHANOL

A study by Northwestern University found that the existing ethanol industry in the United States, which at the time of the study produced

1.5 billion gallons per year, increased net farm income more than $4.5 billion, boosted total employment by 192,000 jobs, improved the balance of trade by $2 billion, added over $450 million to state tax receipts, and resulted in a net federal budget savings of over $3.5 billion.

Not bad, right?

And that was in 1997, when the modern U.S. ethanol industry was in its infancy. Wouldn't it be foolish not to fully support a business that generates these sorts of numbers when it is just starting out—foolish not to nurture the full potential contribution the industry could make to the economy? Not wondering how the industry could perform when it was all grown up would require a terrible lack of imagination.

Technology has come a long way in the eleven years since that study. Most of the rest of the world still doesn't have the growing conditions, of course, that would allow a high-energy crop like sugar to flourish within its borders. In that respect, to go back to our audio analogy, Brazil has invented the shiny new CD and the rest of us are just dancing to the music. We do have something else, however. Imagine, if you will, what those numbers from Northwestern might look like if the United States was making its ethanol from the most efficient and sophisticated feedstock yet known. What might happen if the university took its measure then?

Cellulosic ethanol is an umbrella term for the ethanol that's produced from lignocelluloses. Lignocellulose is a plant's structural material. It is, in fact, what makes up the bulk of the plant—its stalk, leaves, and twigs, where two-thirds of the energy the plant absorbs from the sun is stored. In the cellulosic process, special enzymes are used to break down a plant's fibrous structures and release the energy stored there, within those "tougher" molecules. It allows for the use of the whole plant to create ethanol and ethanol by-products, providing, pound for pound and bushel for bushel, more energy than from any other known production process. The resulting ethanol, which is chemically identical to the ethanol produced from corn or sugar in how it allows a car to perform, has, in addition, a much greater impact in reducing greenhouse gases—up to

85 percent over standard gasoline, a stunning advance over the 17 percent that corn ethanol can claim.

And the price is right too. Depending on the place where it is manufactured, and as of this writing, the production cost for a gallon of corn ethanol is about $2.00. For gasoline, it is 90 cents. For sugar ethanol, 75 cents. For cellulosic ethanol, the cost of producing one gallon is expected to be as low as 69 cents.

That's right. It's *expected* to be.

Cellulosic ethanol is an industry in its infancy. It's the next iteration—the iPod, if you will, according to the audio illustration we've been using; the next big thing in our renewable fuel future.

How does cellulosic ethanol come by all of its advantages and benefits?

Possibly the greatest advantage of the cellulosic process is that it allows us to make fuel from an astonishing array of feedstocks. Sit back and get comfortable, Madam Mayor; this is a long list.

Cellulosic ethanol can be made from switchgrass, a natural, perennial prairie grass. In comparison to corn, it is a crop that almost cultivates itself. It needs less topsoil and less nitrogen to grow heartily, eliminating the need for petroleum-based fertilizers—and the associated water pollution from nitrogen runoff. To the contrary, in fact, switchgrass actually enriches the soil's carbon content, acting as a voracious carbon sink. As an added benefit, one of the by-products of using switchgrass as an ethanol feedstock is cattle feed. And, as is the case for every cellulosic feedstock, the lignin—waste left over after the plant is processed—can be burned to generate the electricity required to transform the plant into fuel, doing away with the need to use fossil fuel in any step of the ethanol manufacturing process and— potentially—producing enough electricity that an excess can be sold to the grid.

Cellulosic ethanol can be made from soybeans, wild ryes, millet, miscanthus, orchard grass, blue joint reed grass, reed canary grass, prairie sand reed, and Jerusalem artichoke, to name but a few. It can also be made from corn *stover*—the waste left in the fields after corn is harvested—or the stalks and stubble of wheat or oats—feedstocks that have the potential to diversify the income streams from fields that are planted with corn or wheat for food. And, speaking of diversity, the cultivation of a spectrum of energy crops like these allows for a biodiversity that focusing on one primary energy crop simply can't provide. Wise management of these crops, and the crop rotation to which they lend themselves, is critical in maintaining the integrity of the soil in which they're grown.

Oh, yes, and the natural grasses we've mentioned in the last paragraph? They provide natural habitats for wildlife.

Cellulosic ethanol can be made from wood waste and from forest residue. The state of Oregon, where forests are abundant, did a study that concluded that by thinning just 2 percent of its forested land a year to provide feedstock for cellulosic production, it could produce nearly 200 million gallons of ethanol. This thinning is necessary, as it keeps a forest healthy. Turning the residue into cellulosic ethanol would pay for the thinning costs—and the state could also save some of the $25 million a year it now spends fighting fires that result when a forest is neglected.

Cellulosic ethanol can be made from paper mill sludge—an opportunity a paper mill operator might well welcome as there is now a fee associated with sludge disposal; the operator could turn a liability into a profit maker by selling the sludge to an ethanol manufacturer. It also can be made from urban greenwaste—lawn clippings and food wastes— which would divert your trash from a landfill and, in the process, according to some estimates, decrease greenhouse gases emanating from landfills by up to 90 percent. Partnering with an ethanol production facility might be an attractive option for a local waste management firm that must currently pay a tipping fee to dispose of the trash it collects on its weekly run through your neighborhood. Cellulosic ethanol can be made from potato residues too, a fact that might tickle your imagination if you live in Idaho.

This list is incomplete, but you get the idea. Take a look around. Opportunities for renewable fuels are all over. When a friend of ours who lives in the northern California wine country read a draft of this chapter, she did just that and suddenly wondered what winemakers did with the stems and leaves of the vines after the grapes were pressed in the fall. "*Grape* ethanol. Think of that!"

The idea of being able to produce a valuable fuel from waste materials like grape stems and potato residue, lawn clippings and paper mill sludge, appeals to the conservationist in all of us. Saving those waste products from taking up precious space in a landfill—and the planet from the methane gas they release while they're decomposing there—will be the hallmark of a truly comprehensive and streamlined plan to solving our greater energy challenge.

But grapes are seasonal, and the paper mill may well decide to put its sludge to use, burning it in its own boilers to generate the electricity it needs to run its machinery and thereby becoming energy self-sufficient. Grape stems and paper sludge will likely turn out to be economical but supplemental feedstocks for the production of cellulosic ethanol, with the bulk of production relying on cultivated energy crops, such as switchgrass, or waste from food crops, like corn stover or wheat stubble. Your next logical questions are going to be: How much land are all of these energy crops going to take up? Do we have enough of it? And, if we do, will the energy crops be profitable for farmers to cultivate? Here is a story that answers those questions neatly.

In the United States, in 2006, Congress passed an energy act calling for the doubling of the country's biofuel production, to 7.5 billion gallons annually, by the year 2012. Judging by the estimated potential for energy crop cultivation, that's not an ambitious goal. According the National Resource Defense Council, the country could produce *165* billion gallons of ethanol from existing cropland—and still be able to meet other agricultural needs. The state of South Dakota alone has enough available land to produce 3.429 thousand barrels of cellulosic ethanol per day, enough to meet 30 percent of the country's total motor fuel needs—and, incidentally, make that one state the sole rival to Saudi Arabia as America's biggest energy supplier.

What would this sort of productivity mean for rural America? The cultivation of dedicated energy crops has been projected to boost the rural gross domestic product in the United States anywhere from $5 to $50 billion.

Five to $50 billion?

How do we make that happen?

Leadership. Good local, regional, and national leadership is the key.

In November 2007, Montana's governor Brian Schweitzer, along with U.S. Senators Max Baucus and Jon Tester, representatives from state agriculture, and two companies, Targeted Growth, Inc. and Green Earth Fuels, announced a joint venture called Sustainable Oils, Inc. to produce and market up to 100 million gallons of camelina-based biodiesel by 2010, the single largest U.S. contract for a unique biodiesel feedstock. Nearly all of the camelina will be grown in Montana.

"What's most exciting about this new project is that Montana is going to be part of the energy solution," Senator Baucus, chairman of the Senate Finance Committee, said of the announcement. "The fertile fields of Big Sky Country will be on the cutting edge of a bright energy future for America. And not only are we developing new cleaner energy sources, we are creating jobs and boosting Montana's economy too."

"If I had a bit more time on my farm," joked Senator Tester, the sponsor of legislation to provide Montana farmers with crop insurance for camelina, "I'd plant some camelina myself."

Well, all right for Montana! Now, how do the *rest* of us make it happen? Keep reading.

7

CREATING A RENEWABLE INDUSTRY

ALL ACTIVITY REQUIRES THE EXPENDITURE OF ENERGY. WHEN WE TAKE OUR terrier, Liberty, out to the park to play fetch with her favorite blue India rubber ball, we're spending human and animal energy on an activity that gives all of us a great deal of pleasure. We produce the energy that allows us to play a good game of fetch through eating and sleeping—and we *pay* for the bulk of that energy every week at the grocery store. The question that's probably crossed your mind is how are we going to pay for the energy we need to drive our economic engines as we transition into the coming renewable age? Who's paying to develop and deploy all of those wind farms and geothermal plants now, and who will pay to construct them on the scale we require to keep pace with both our growing demand for energy and the earth's need for that energy to be green?

Fueling our bodies with food, burning up the calories we take in, and then replenishing our energy stores with another meal is the most basic energy cycle of all. The economic energy cycle might seem to be less fundamental, but it shares the characteristic of momentum with our human and animal cycles. One of Liberty's favorite ways to spend a weekend afternoon is taking a long hike in the woods. When we return home from these treks we're all ravenous, eager to dig into a bowl of fruit or a bowl of kibble—depending on our personal tastes—because, in expending so much energy, we've depleted our stores and created the need for more fuel to carry us through to the next activity.

So goes the cycle of our economic engines too: Electrical or mechanical energy is required to accomplish all of the myriad tasks that enable us to be participants in the marketplace; in spending that energy by producing ideas, goods, and services, we create the need for more energy to take us through the next business day. In fact, if we have spent our economic energy wisely—employing the most efficient means to produce our best-quality ideas, goods, and services that shoppers in the marketplace will really want or undoubtedly require—we create the need for even more energy to take us through to the next activity than we needed the first time around because the demand for what we produce has grown. In order to satisfy the customers now clamoring for our product, we find we need more office space, an expanded manufacturing facility, more computers and phone lines and other business tools to equip the additional employees we have to hire. And that, of course—successfully growing our business—was the whole point in the first place.

We have a friend who lives in a very small town. Several years ago she opened a very small fine-dining restaurant. It wasn't the only restaurant in town, certainly, but her superb food, comfortable décor, and warm enthusiasm with which she greeted her patrons and made them feel welcome soon put her at the top of the heap. She mentioned to some of her customers and friends that she was thinking about expanding her business. She needed to add more tables to accommodate all of the people who wanted to eat at her place, and she needed some special equipment—a salamander and a bread oven—so that she could enlarge the menu she was able to offer to her guests.

News of her idea traveled quickly, as news will in a small town, and before long a representative from her community's economic development committee came to call. He offered help in structuring her plans for the expansion of her business and in applying for the low-interest loans and even a few grants available through state and local agencies to assist businesspeople in growing small enterprises. Our friend was pleased if a little surprised by his visit, but we weren't surprised at all that the economic development committee had come to court her. She had a proven track record of success to show that she was very good at what she did, and she turned a profit at it. The folks who sat on the committee knew that in supporting the growth of her business, the community could benefit in the long term through job creation and an increased tax base—not to mention that in recruiting industry to the town, assets like good schools, well-maintained public parks, and a nice restaurant or two are real plusses.

One of the government's most fundamental roles is to facilitate economic growth. Whether it's a local development committee helping a small restaurateur find ways to finance the growth of her business or a federal government enacting legislation that helps to sustain a much larger industry, government agencies often intervene to create a favorable business climate so the community as a whole will benefit.

The transition to renewable energy will create a global industry on a grand scale. That's a sweeping statement, but we make it with a full understanding of the unique position that the renewable energy industry holds in our future.

And just what is that position?

The transition to renewable sources of power and fuel is necessary for several profound reasons. Let's revisit them briefly.

1. We are likely running out of at least one of the fossil fuels that has helped those of us in the developed world attain a standard of living unmatched at any other time in history and that is now helping those in developing nations to join the global marketplace and prosper. But even if we weren't running out of oil—as some claim—the transition to a renewable energy economy would still be a prudent one, because it's improbable that oil's high price will decrease. The reason for that is the soaring costs of finding the reserves of oil that are left in the world and extracting them. For example, the Canadian Tar Sands, located in Alberta, Canada, have been touted as a vast, untapped supply of oil. But recovering oil from tar sands is a more involved, and therefore more expensive, proposition than conventional oil recovery. Tar sands must be mined, using strip mining or open pit techniques, and the bitumen— a viscous, black oil—that is extracted from them must be separated from the clay, water, and sand that makes up the tar sands. Then the bitumen must be upgraded and refined and, because it is so heavy, diluted with lighter hydrocarbons so that it can be transported through pipelines. About two tons of sand are needed to produce one barrel of oil. This complex recovery and refining process, coupled with the environmental impacts of mining tar sands, make bitumen's cost to the consumer unattractive, to say the least, as a long-term energy strategy. The bottom line is that whether or not you believe the world oil supply is dwindling, there are steep economic and

environmental costs associated with continuing to extract it from the Earth. These costs make it clear that it is time to put all of our energy options on the table and choose among them intelligently.

2. We know full well via repeated reports from the world's leading scientists that the consequences of continuing to use fossil fuels are grave—the practice is, even now, creating natural and economic catastrophes, from fires in southern California to melting ice at the South Pole.

3. Science and technology have provided us with every tool we need to wean ourselves from fossil fuels and avert these catastrophic natural and economic events.

Now, here's the unique part: These tools—these renewable resources at our disposal for the asking—can be made to work for us profitably. Transitioning to renewable energy is something we don't just *have* to do; it's something we can *want* to do. The means to solve our energy challenge will enrich the earth at the same time that it creates an enriching new industry. The switch to renewables will give energy suppliers, energy distributors, and energy users—all of us—the opportunity to do well by doing good.

Renewable energy sources already have a proven track record for being good investments. Wal-Mart, for example, has famously gone green. According to the company itself, in the process of adopting conservation measures, such as working with its suppliers to reduce the amount of packaging materials for their merchandise and adding solar and wind energy systems to their shopping complexes, Wal-Mart estimates that it has prevented 667,000 metric tons of carbon dioxide (CO_2) from entering the atmosphere—and saved $10.98 billion on the corporation's bottom line.

BP, the energy giant that launched the $8 billion renewable program we talked about in chapter 4, has reduced its own greenhouse gas emissions to 10 percent below its 1990 level and, as a result, added $650 million in value to the company.

Michael took a very informed risk about three years ago, when he acquired that sparkling solar field in the Mojave Desert. A $50 million upgrade later, including replacing the heat transfer fluid in the field's pipes with an oil that burned hotter and so increased the field's efficiency by 30 percent, he's just sold that same solar field that was such a risk in 2005 for four times the amount that was invested.

The merchants on U Street are enjoying clean energy for less money than they would have had to pay for the dirty stuff.

We already know that renewable energy technologies are, from a financial perspective alone, a very good bet. Also, we know that as the technology comes more and more into common use, its price will decrease. Anyone who bought, say, an iPhone when the device first hit the market will tell you, however grudgingly, that as technology becomes conventional, its price comes down. And as the cost of technology decreases, it becomes even more profitable for suppliers and more affordable to users.

Why, then, given the enormous payoffs that renewable energy promises for the global community, haven't government agencies the world over come courting? Why, for instance, isn't the United States, right at this very moment, aggressively empowering farmers to cultivate the crops and manufacture the cellulosic ethanol that would add $5 to $50 billion to its rural gross domestic product? The development and deployment of renewable technologies is being left to forward-thinking corporations, such as Google, which, in December 2007, formed an internal research group and hired a team of engineers whose sole goal it is to produce 1 gigawatt of renewable energy capacity—enough to power all of San Francisco, near where the company is headquartered—that is cheaper to use than coal. Or to volunteers, like the 600-member Energy Club at MIT that has organized twenty-four college research teams from around the world to build an efficient plug-in electric hybrid car within three years. Or to private investors who—as we'll see in the next section of this chapter—labor under some hard disadvantages. Why aren't governments supporting these private initiatives in every way possible to effect the transition that's required with all due haste? Why does the economic energy cycle seem so often to sputter and stall in the case of renewable power and fuels?

That's what this chapter is all about—the ways in which government policies are actually standing in the way of the development of renewable technologies, and the ways in which that can, and is, starting to change.

THE BIG LIE

Let's say for the sake of argument that you are in the market for a new sofa. You want a large, soft, comfortable sofa and you want it in the color green. In the store there are two green sofas. The first one is a hunter green that might go all right with the rest of your décor, but the fabric is a little scratchy and the cushions aren't as plump as you'd like them to be. Still, it's only $800. Right next to it, however, is the perfect deep sage green sofa with linen slipcovers and feather stuffing. It's also only $800. Your choice seems clear. And then the salesman walks over to take your order and informs you that the manufacturer of the scratchy, thin hunter green sofa is offering a $200 rebate. Depending on how much price is a factor to you in this purchase—or maybe on how invested you are in the look you want for your refurnished living room—the rebate has made your choice more difficult. In any case, it has distorted the true value of the sofas.

There is a widespread conceit that renewable energy is more expensive than fossil fuel–generated energy. That's simply not true. What is true, at this point in time, is that most renewable energy is not yet cost competitive with its fossil fuel counterpart. The reason for this is that fossil fuel industries are subsidized—heavily subsidized—by governments around the world. This distorts the value of every energy source, making traditional energy seem cheap and clean energy costlier than, in reality, it is or, indeed, needs to be.

Before we delve into the subject of subsidies, there is one important point we need you to keep in mind: While we certainly are going to make the case in the upcoming pages for increased subsidy support for renewable energy technologies, we are in no way advocating that the fossil fuel industry be unceremoniously stripped of its subsidies in order to accomplish this. We may all prefer clean energy but the hard truth remains that fossil fuel energy is going to be a part of our lives for likely decades to come. A lion's share of our electricity is right now produced by coal; in the great majority of cases our cars run on petroleum products. If we stripped fossil fuel subsidies we would be stripping down our standards of living—and the economies that support them—to the bone. Rather, what we are advocating is a more reasonable disbursement of government funds—a more equitable distribution that allows fossil fuel prices to remain stable while the development and deployment of renewable technologies is financed and these new energy sources take their rightful place in our twenty-first century energy mix.

A subsidy is defined by the World Trade Organization as "a financial contribution by a government or any public body . . . or . . . any form of income or price support [whereby] a benefit is thereby conferred." Such benefits are bestowed on oil, coal, and natural gas industries through a variety of avenues.

Some subsidies take the form of preferential insurance or tax rates well below the national corporate norm. Some subsidies benefit a particular sector of the energy market, such as government-financed oil-related research and development programs. Other subsidies are offered as depletion allowances, tax-free construction bonds, or below-cost loans with lenient repayment schedules. Consumption subsidies, which are government measures that seek to directly provide the energy end user with a price below what would prevail in a truly competitive market, are common in Indonesia and Venezuela. In Saudi Arabia, consumption of oil is so heavily subsidized that its people have come to think of cheap energy as a right of citizenship.

Governments provide for subsides to energy companies for one very logical reason: Affordable energy is crucial to sustained economic well-being. Subsidy programs at their best help to insure jobs, growth, and the lubrication necessary to keep a nation's economic engines churning in a competitive world market. But the way in which subsidy programs currently function—and the types of fuels they benefit—were modeled around old ways of thinking about traditional power, and, too often, they don't even directly benefit the consumer. It's this old model that is a tremendous roadblock in the way of the development and deployment of clean energy technologies because it keeps the price of oil, coal, and natural gas artificially below their market value and, at the same time, keeps the price of renewables, which are, in the main, largely unsupported, high in comparison.

Though studies have attempted to illustrate the ways in which fossil fuel subsidies are conferred worldwide and to set a dollar value to them, the scale of such an undertaking is daunting, hampered by the sheer volume of data, the complexity of laws and regulations that affect the subsidy disbursement differently in each country, and the difficulty in gathering a complete set of information from regimes reluctant to participate transparently. One study, carried out by the World Bank in 1992, that attempted to quantify worldwide subsidies to the oil industry concluded that they stood somewhere around $230 billion.

More recent figures, however, estimate that in the United States alone, $200 billion of the federal budget is set aside annually to satisfy fossil fuel subsidies. Compare that $200 billion to the $33 billion currently allotted in the U.S. federal budget for the development and deployment of new renewable power and fuels, and you'll see how this imbalance distorts the market value of renewables—and impedes the progress of a necessary new industry. It would be a different world if these subsidy numbers were reversed.

Government policies favoring traditional fuels greatly reduce their costs in relation to the renewables that must replace them. But they also do something else. These policies provide the means to finance the expansion of fossil fuel exploration and facility construction. This, in turn, keeps the price of these sources of energy artificially low. It is these artificial prices that work against the conservation measures we need to implement in order to sustain our economies by removing the incentive for people to use less fossil fuels.

The continuation of policies like these, made in the past under a now outdated set of assumptions about how to power our world, block the way to realizing our energy future.

There is one additional—and vast—way in which fossil fuel industries benefit from government subsidies, a hidden way that is often acknowledged but rarely taken into real dollars-and-cents consideration when subsidies are analyzed. This is the toll taken on the environment by polluting industries.

The associated costs of pollution are often borne by the government and include a range of categories from environmental mitigation necessitated by oil spills, to the health care of people suffering from pollution-related illness—diseases like asthma, cancers, and immune disorders.

Individual citizens and the governments they live under are seeking out ways within the existing structures of their federal laws and budgets to realize the timely development and rapid deployment of renewable technologies. The focus has turned to requiring that polluting companies take responsibility for the pollution they generate as a way of paying for the biofuel infrastructure and solar fields and other research and construction that must happen for us to make our way to our energy future.

In the balance of this chapter we present the three main ways that individuals, corporations, and governments are now working to erase the funding disparities between fossil fuel energy sources and renewable energy sources. Carbon offsets, renewable energy certificates, and carbon cap-and-trade programs are all ways of paying for pollution. In every case, the bulk of the monies these sorts of programs generate is, or should be, directed to programs that support the expansion of renewable technology research and use.

CARBON OFFSETS

Carbon-offset programs work through firms that collect a voluntary fee from individuals or organizations for engaging in carbon-producing activities and, in turn, invest those fees in projects that will offset, or balance, the carbon that is produced with some sort of mitigation measure. Taking a flight to get to your vacation destination, or putting up a display of holiday lights, or simply driving your car are all activities that emit carbon. You can offset the carbon you generate during each of them by paying a fee to a carbon-offsetting firm, which will direct your money to a tree-planting initiative in Ecuador, the construction of a wind farm in East Africa, or a green school solar-roof program on the east side of your hometown.

In theory, this is a fine idea. Though this method of carbon offsetting offers at best incremental solutions to the larger problem of funding our energy future—typically one carbon offset can be purchased for around $20, and some can be purchased for as little as $5—we like the concept for how it raises each person's awareness of how his or her personal lifestyle activities impact our energy environment. It's also empowering for people to be able to make such a direct contribution to the earth's welfare.

Not everyone, however, wholly embraces carbon-offset programs. The English environmentalist and writer George Monbiot has likened carbon-offset programs to the medieval practice of buying indulgences from the church whereby sins would be forgiven without the sinner having to go to the trouble of either repenting or ceasing to sin. In other words, even though a price is paid toward the mitigation of the pollution, the pollution still exists. But, let's face it, we live in a world where business can frequently take us to distant cities and family whose company we yearn for can be far-flung, so ceasing to travel is unrealistic. Making use of carbon offsets to balance the carbon we create personally can be a meaningful step.

Carbon offsets have also, however, come under scrutiny for misuse or unwise use of the funds they collect. The most famous case, probably, was when the Academy of Motion Picture Arts and Sciences paid a firm called TerraPass to offset the carbon generated by its 2006 Oscar telecast, and it was discovered that the money went to a methane recapture project in Arkansas. This project predated the TerraPass offsets and was, in any case, legally mandated; the owners would have had to move forward with the mitigation with or without the academy's money. Carbon offsets, ideally and, we think, legally, should be required to fund mitigation efforts that are in *addition* to those programs either already in place or required by law. Otherwise they don't serve as a tool to mitigate carbon emissions caused by their buyers but go only toward the cost of mitigating carbon emissions that were already accounted for.

One friend of ours had a rather different objection to carbon offsets. Reacting to the voluntary nature of such programs, and what she felt was a poor substitute for a government's committed and coordinated effort to rein in a global energy crisis, she told us she thought carbon-offset programs were a little bit like asking the Air Force to hold a bake sale to buy a bomber.

Still, the theory behind carbon offsetting is a sound one. Empowering private people to help solve a public problem is a credible way to engage both their spirits and their pocketbooks. Think of the war bonds that were sold in the 1940s. It's also, surely, not a bad way at all to raise significant funds, however incrementally. Ask any political candidate how much of their campaign was funded by small donations; you'll likely be astonished at how quickly these seemingly small amounts can add up.

What is needed here is a credible way to monitor the use of the fees paid to carbon-offset firms. In order for this method of funding the renewable future to have an impact and become a conventional and convenient way for people to take part in realizing the renewable age, they have to feel comfortable that their money is being well spent on well-thought-out projects. A licensing arrangement for carbon-offset companies and an official review process for the projects to which a company proposes to direct its collected fees would go a long way toward achieving this end.

RENEWABLE ENERGY CERTIFICATES

Renewable energy certificates (RECs) go by a lot of different names, including renewable energy credits, tradable renewable certificates (TRCs), and green tags. They can be purchased by institutions as large as the Environmental Protection Agency; by businesses as small as your hometown diner, bank branch, or yoga studio; and by energy consumers as small as your next-door neighbor who lives alone in a one-bedroom house with three cats. They all have one thing in common: What they represent is a specific *attribute* of energy.

One attribute of energy is, of course, the heat or electricity or motion that it produces for our use. This is the attribute conventionally sold by power suppliers and distributors.

But another attribute of energy, exclusive to green sources of energy, is the positive impact its use has on the environment through the reduction of greenhouse gases and other pollutants. It is this attribute—legally defined as the property rights to the environmental benefits of renewable energy—that is bought, sold, and traded through the medium of green tags.

Let's make this point even more clear. When you pay your electric bill at the end of the month, you are paying for heat, or the juice to run your toaster oven. If some of the energy you use to run your toaster oven has been generated at a renewable facility—a geothermal plant, for instance—there is a second commodity for sale: the energy's "greenness," or its environmentally beneficial component, and this is a commodity that can be purchased separately. I can theoretically own the environmental attribute of the energy you use to run your toaster oven by buying a green tag.

How do green tags work? A renewable energy provider, such as a solar field or a geothermal plant, is credited with one green tag for every 1,000 kilowatt-hours of electricity that it produces. A certifying agency authorizes the tag; usually it's a state agency in the United States, though state programs are increasingly being tracked through regional oversight boards set up by a coalition of states. Each tag is stamped with an identification number. This unique number assures that the tag isn't double counted. When the green tag has received its ID number and the green energy has been fed into the grid, the green tag can be sold on the open market.

Now, what exactly do we mean by "double counted"?

The concept of "additionality" plays a key role in making green tags effective. This means that, in most places where REC programs are in place, green tags can't be used to fulfill renewable energy standards that are legally mandated. If, for instance, a state has passed a law that calls for energy providers to produce a minimum of 10 percent of the electricity they generate with renewable resources, the provider can't use a green tag to fulfill a portion of that mandated percentage. Green tags are meant to be used to *increase* the amount of renewable energy used by going above and beyond whatever green is required by law. Because green tags are a voluntary purchase for the energy user— and as the purchaser pays a premium to own the environmental attributes of the energy, which often but may not always result in a decrease in his or her monthly electric bill, as it did for the U Street merchants—we see the need to avoid double counting as a core matter of consumer protection.

A national registry of REC ID numbers, which does not currently exist in the United States, would enhance the ability to verify that the value of the environmental attributes purchased with each green tag is counted once, and only once.

Green tags provide for the growth of renewable energy technologies in two very important ways. First, being able to sell the environmental benefits of green energy as a separate commodity from the heat or electricity they generate creates a second income stream for green power providers. They are then able to reinvest these additional funds to build more green power plants—or to cover the extra costs that are, at this time, associated with green energy.

We have talked about the heavy subsidies that oil companies and other fossil fuel producers receive from governments. In the United States, renewable power is currently supported by meager tax incentives, some as low as .019 cents per kilowatt-hour generated. The supplemental funds created by green tags are an important source of another kind of green power for renewable energy producers to use to sustain and expand their operations.

The second way in which green tags are important to reaching the goal of a renewably powered society is that when they are purchased, the energy they represent must be fed into the grid by mandate. If you, for example, purchase one green tag, which, as we've said, generally represents 1,000 kilowatt-hours, or a little more than the average household uses in one month, your energy provider must purchase enough renewable energy to feed your home with electricity for one month.

Now, in purchasing one single green tag as a private citizen, you probably won't enjoy a reduced monthly electricity bill. In order to do that, you will most likely have to do what the U Street merchants did: Form a neighborhood coalition of home and business owners who can pool their electricity use with yours and then use a broker to negotiate a volume discount with your energy provider, based on your group's total monthly usage.

For the purpose of our example, however, let's stick for a moment with just that one green tag, which represents 100 percent green power to your home for a one-month period. What this means, in practice, is that you have displaced your traditional energy source for one month. The fossil fuel that would have otherwise been burned to provide your home with electricity went unused, and the greenhouse gas emissions associated with that amount of coal or natural gas went unreleased.

Now, the more green tags you and your neighbors buy, the more money you will likely save as a group, but, in addition, the more fossil fuel you will displace. Remember, electricity is generated on an as-needed basis, fresh to you as you use it. A supplier can't throw a ton of coal into the boiler to manufacture electricity and stockpile it. Remember too that once a green tag is sold, the renewable energy associated with it must by law be fed into the grid. Your supplier must go out and find a wind farm or a geothermal plant or a solar field to provide the 1,000 kilowatt-hours of power to serve up to you that month, and every month thereafter for twenty-four or thirty-six months or however long your coalition has contracted for its green energy.

The more green tags that are sold, the more green energy is used, and the more fossil fuels are displaced, and the more greenhouse gases are saved from being emitted into the atmosphere, and the more the cost of renewable energy comes down, and the more affordable green energy becomes in relation to traditional energy. The more people buy green tags, the more green energy is used, and the more green power plants can be built to meet the demand, and the more green power plants that fill the demand the less need for coal-fired plants . . .

You see how it can work.

Our daughter likes to "borrow" my BlackBerry, or her father's, and play a round or two of a game called Brickbreaker. In it the player uses a small paddle to whack a very tiny ball into a wall of brick. As it takes each brick four or five strikes with the ball to crumble, the wall is slowly chipped away. I've watched her play a few times, and suddenly it hit me: That very tiny ball slowly but surely chipping away at the daunting wall of solid brick is like a green tag; every time one hits a coal-fired plant, the closer coal comes to crumbling.

Green tags are a powerful tool for energy consumers to use to grow the demand for clean energy and to help eventually displace fossil fuels altogether.

CARBON CAP-AND-TRADE PROGRAMS

What do Alan Greenspan and Greenpeace have in common? They both believe that the time has come to start making polluters pay for the environmental damage they cause. New York's mayor, Michael Bloomberg, put it this way: "We have to stop ignoring the laws of economics. As long as greenhouse gas pollution is free, it will be abundant. If we want to reduce it, there has to be a cost for producing it."

The mayor went on to add: "If we're serious about climate change, the question is not whether we should put a value on greenhouse gas pollution, but how we should do it."

So, how should we do it?

In the United States, two ways of putting a value on carbon emissions are being weighed. The first is a carbon tax, an approach advocated by Mayor Bloomberg as well as Al Gore, among others.

A carbon tax would indeed seem to be the most direct approach. Every metric ton of CO_2 emitted would be worth a certain set dollar amount. It would be a simple matter then to multiply the tons of carbon produced by a polluting industry by the set dollar figure to know how much the polluting activity was going to cost it. Certainly a carbon tax gets right to the point of raising the cost of engaging in unacceptable behavior and getting the perpetrator to pay. Any of us who are parents understand well the efficacy of punishing bad behavior. Sweden, Finland, the Netherlands, and Norway have had a carbon tax since the early 1990s, and it has worked well. It's a transparent method, and it's easy to administer.

The drawback to the carbon tax approach is that it's a price instrument. That is, it fixes a price on carbon that, in theory, should make it more cost effective for an industry to reduce its emissions than to continue to pay for them. The emissions, however—which are the real problem here—aren't what is addressed in a carbon tax scenario. No goal is set for the actual, real-world reduction of the amount of CO_2 in the atmosphere. In practice, it might well turn out to be more convenient for a company to incorporate a fixed price for polluting into its operating costs than it is to fix the pollution problem.

A second way of putting a value on carbon emissions is to implement a cap-and-trade program. This method directly impacts the amount of carbon an industry produces. It's a quantity instrument in that it sets a cap—a fixed upper limit—on the amount of emissions permitted rather than a dollar amount for, theoretically, an unlimited quantity of emissions. And that's the hook: the *mandatory* cap that provides the standard by which environmental mitigation progress is measured, and from which cap-and-trade derives its market value.

Under this system, licenses, or credits, would be issued to companies that would allow them to produce a set amount of pollution—usually 1 ton of CO_2 per credit. If they implement conservation measures, generate their electricity from renewable sources, and/or install equipment that enables them to produce less emissions than they were allotted, they receive credit for the amount of pollution they did not produce. These credits can then be sold for a price set in the open market, or they can be banked for future use, at the company's discretion. They are, after all, what the company has earned in return for its pollution mitigation investment.

If, however, a company exceeds its cap, it can buy the credits it needs to meet its mandated emissions goal.

As a parent, you can liken the cap-and-trade system to the tactic of rewarding good behavior; rather than doling out consistent punishment for messing up, you offer rewards for cleaning up an act within the parameters you've set.

There are drawbacks to the cap-and-trade system too, of course. Nothing's perfect. A cap-and-trade system would be more complex to administer—What body would issue the licenses? On what criteria? And how would carbon produced by the licensees be monitored?—and with that complexity would come higher administrative costs.

In addition, the system would need stringent mechanisms to keep it from being vulnerable to fraud; carbon credits are *money*, after all. The penalties for fraud would need to be severe to ensure that the program remains effective in its purpose of carbon reduction. At the end of every compliance period, each source of pollution must be held accountable for holding a number of credits equal to the tons of CO_2 emitted for that period, for measuring its emissions accurately, and for reporting them with full transparency.

Also, a certain uncertainty is inherent in a cap-and-trade system. An industry's emissions will vary according to economic activity, and the market price of the carbon credits could rise sharply when emissions increase, seasonally or during inclement weather—a hot spell, perhaps, when a power supplier has to rely on a coal-fired plant to provide the extra electricity its customers want for their air conditioners.

Still, we like the cap-and-trade approach to carbon emission reductions for several pretty powerful reasons. The first is that carbon emissions are not a *local* problem. Carbon that is released into the air in Kansas City can end up affecting the Black Forest in Germany. Emissions that originate in Beijing can wander of their own will to the Netherlands. Carbon emissions are a global problem that demands a global solution, and this solution requires a joining together with the international community to place caps—fixed upper limits—on the amount of pollution we will tolerate on our planet. In June 2007 the G8 Climate Change Roundtable, formed in 2005 by twenty-three leading global businesses to formulate climate change strategy, called for a long-term international policy framework within which to curtail the CO_2 problem. Members of the group underscored the need to work together, and with the leaders of countries with emerging economies, such as China and India, to develop international mechanisms to combat the crisis. In fact, a landmark deal was reached at that conference to set a goal to cut carbon gases worldwide by 50 percent by 2050. An agreement to cap-and-trade is a start.

Moreover, a global cap on emissions handily addresses a situation that has the potential to become problematic under country-by-country cap-and-trade systems: *leakage*. "Leakage" is the term for what happens when a company seeks to avoid its responsibility for the pollution it generates in its home country by moving its operations to a location where the cap is unrecognized and it can pollute at will. A globally united front on CO_2 emissions removes this disincentive.

We like a cap-and-trade approach for one more solid reason: We've got proof that it can work very well. The cap-and-trade concept was developed in the United States in the 1990s, as a program within the 1990 Clean Air Act Amendments, specifically to address sulfur dioxide emissions—acid rain. Before the cap-and-trade program was launched, it was estimated that the cost to mitigate the sulfur dioxide emissions in the United States could be anywhere from $3 billion to $25 billion per year. But by instituting this market-based system that allowed industry the flexibility to meet its own mitigation targets as efficiently and cost effectively as possible within its own set of operating conditions, and by providing companies with a direct, tangible financial incentive to comply with federal mandates, in the first two years the cost of mitigation was just under $800,000, far below projections. The acid rain cap program, in fact, achieved not only 100 percent compliance, but sulfur dioxide emissions were reduced 22 percent *below* the mandated level.

The Environmental Defense Fund heralds the sulfur cap-and-trade program as "unprecedented environmental protection at unmatched cost efficiency." *The Economist* magazine labeled it, in 2002, "probably the greatest green success story of the past decade."

The healthy carbon cap-and-trade program already in action in the European Union (EU) further puts the lie to the idea that a cap on emissions is a cap on economic growth. According to a report from the Carbon Finance Unit of the World Bank, carbon trading increased forty-fold in 2005, from 2004, since carbon became a commodity with value. Developing countries also have begun to participate in the EU market, bringing, in the words of the report's authors, an influx of capital, innovation, and "real emission reductions to the table."

In reevaluating how we use and manufacture energy, we also have to reevaluate how we *pay* for energy. Science has laid out the tab for continuing to use fossil fuels, and we fully grasp the gravity. But in order to transition into the new energy age, someone actually has to build geothermal plants and photovoltaic panels and cellulosic refineries that will produce the new power and fuels. And that construction is going to mean someone is going to have to spend money.

How do you want to pay for it?

We believe the answer has four parts:

1. Reallocate some of the federal subsidies that now benefit oil companies to renewable energy suppliers and distributors.
2. Encourage individuals to offset their carbon use through reputable carbon-offset firms.
3. Expand the opportunities for citizens from every country to participate in REC programs.
4. Give industry an incentive to invest in emissions reduction by rewarding pollution controls with valuable, tradable carbon credits.

The final part of this formula, of course, is to direct the funds realized through these programs to producers and distributors of renewable power and fuel. By using these funds to cover what are now higher-than-average capital costs, firms can keep producing green energy, new and expanded renewable energy facilities can be built, and new technology can be investigated, developed, and commercialized to take us to the next generation of power "cleanliness" and cost-effectiveness.

These are the ways in which the renewable energy cycle is going to be kick-started. These are the ways to create an industry and the jobs that go with it. They're big steps, but ones we must take to make our way to our energy future.

8

TECHNOLOGIES THAT DON'T EXIST YET

A "BLACK BOX" IS WHAT TECHNOLOGISTS, REGULATORS, AND INVESTORS CALL an unknown new invention. It's a package that lands on their desk, and no one but the inventor knows what's inside, or how it works.

Or *if* it works.

Jeff Voorhis of the Texas Commission for Environmental Quality has opened up a whole slew of black boxes. He explains the volume that passes through his office this way: "In Texas we've got the most air pollution, the most hazardous waste, but we've got bigger innovations too. Everything in Texas is bigger."

By the time a black box hits a desk in a regulatory agency like the one Jeff works for, inventors usually are confident enough that the outcome of their experiments can be reliably duplicated. They're ready to take the next step, and that's Jeff's job—to help prove the technology works. This means, of course, first assessing the project and deciding if the science behind it is viable and has valuable applications. If it does, then the lengthy, expensive, and often frustrating process of shepherding the invention through approved and appropriate testing labs to get third-party verification begins. This independent third-party verification is what will give the new invention the rock-solid credibility it requires before it can move on to the next phases of its life: manufacture, marketing, and finding its niche in people's everyday lives.

In our formula, if you'll remember, we gave equal weight to three different components that, taken together, will secure our energy future: renewable energy, increased energy conservation, and energy technologies that don't exist yet. This last—opening up the black boxes, seeing and assessing and testing the mysteries inside, discovering if one of them can expand on existing technology or if it's something so new it's a technology all to itself, and how it can improve lives—is one of the most exciting parts of building the renewable age. Michael and I sometimes envy jobs, like Jeff's, that have the potential to put a person so directly, every day, at the intersection of the future.

In this chapter we're going to talk about three different new and explosive ideas. One of them has to do with improving the efficiency of gasoline-powered cars, a simple gadget that could bridge our gasoline present to our biofuel future by revolutionizing the internal combustion engine. Another has to do with a waste disposal technology so radical that it could redefine our language by eliminating the very word "waste" from our energy vocabulary. Both of these technologies, as it happens, originate in the big state of Texas.

The third innovation we want to tell you about is happening a world away, in Africa, where a potential new biofuel crop presents what are just crazy-optimistic opportunities for both a brand-new energy source and a broad solution to that continent's crushing poverty. But the care with which we will have to go about cultivating this crop underpins how large our vision of the future has got to be if we're going to get things right this time and the respect with which we must approach our interactions with what the earth offers us.

HOT CARS

Lloyd Spragins is the CEO of Fuelco in Austin, Texas, a company that focuses on the development of emerging technology. He is very particular about the language he uses when he talks about energy. Energy *conservation* is what happens when you turn your thermostat down to 68 degrees from 72 or drive your car at 55 rather than at 75 miles per hour. It's an action you personally take to conserve actual, physical resources.

Energy *efficiency*, however, is improving the ways in which we use existing technologies. It's what the Energy Star program is all about—using appliances with upgraded technologies to save on the amount of electrical watts the appliance must be fed in order to perform its task. In the building trades, says Lloyd, who started out in the construction business, appliances that adhere to Energy Star guidelines are now simply, frequently routinely, incorporated in the building process. Lloyd calls it the "silent revolution." Sometimes these appliances can be a bit more expensive than their energy-inefficient counterparts, and the benefits of spending a little extra up front to save money on energy in the long term might be a hard sell.

"People tend to do what they did yesterday," Lloyd says. "But when you use less energy, there's more money to spend on other things." When you put it that way, people come around.

Energy efficiency is, in a nutshell, increasing productivity and profitability by doing the same amount of work with less energy. Lloyd Spragins has lately turned his attention to improving the efficiency of the internal combustion engine. What motivates him are a few shocking things he knows about cars.

Since 1975, when in the wake of the Arab oil embargo car efficiency standards became a hot topic, carmakers have made an annual gain of only about 1 percent in overall car performance efficiency. They've done this by using lighter materials and more aerodynamic designs, and by modifying the engines. Lloyd proposes that it's time for a much bigger leap.

Did you know that only 30 percent of the fuel you put into your car is actually used to produce mechanical energy? The rest—70 percent—goes up in emissions from your tailpipe. And did you know that 80 percent of all pollution from running a car happens in your garage or driveway? After a car is warmed up, the catalytic converter uses fuel capably, cooking up the molecules. It's at the cold start, when there is no heat to burn the fuel, that the efficiency is lost.

Lloyd and his colleagues are now working on a little gadget that will make a car's engine burn hotter, faster, squeezing as much energy as possible out of every molecule of gasoline and putting it to work for you, saving tailpipe emissions and your gasoline dollar. But that's not the whole beauty of this device. You won't have to buy a new car before you can own one and enjoy its gas savings. Carmakers won't need to make any changes in their production lines either. This little device can be retrofitted to the car you drive now.

Well, then, you ask, if this device is so terrific, why isn't it on the market? Why can't I get one?

Lloyd retired from the construction business and now spends his energies trying to find the people, companies, and products that make things work better. He looks in what you might think of as improbable places. New technology, like the fuel efficiency device, he says, isn't coming out of academia or large corporations. It's coming out of people's garages. It's coming from the same places where Hewlett-Packard started. It's coming from the workbenches where ordinary folks, motivated by a drive to help solve our energy challenge, are innovating in amazing ways. What keeps their innovations from getting out of the garage and into the marketplace is often money.

A lot of myths are intertwined with the history of energy. One of the most famous ones is probably that Rudolph Diesel, shortly after he presented his breakthrough engine at the Paris Exhibition of 1898, was murdered by oil company operatives who didn't want his biofuel invention taking away their business and eating up their profits. But believe it or not, oil companies are in favor of gas-saving measures. If we can double fuel efficiency, they can double their reserves. There have been, says Lloyd, likely a thousand patent applications dating back to 1932 from people trying to invent the same sort of device as the one he and his team are now developing. It's safe to say that 10,000 of the world's top engineers have spent time working on a similar concept: how to get a car engine to burn hot faster in order to more effectively use the fuel that's fed into it. What has impeded such devices from getting to a shelf at your local auto supply store isn't oil company greed or stealth. A lack of money has kept them languishing in patent office archives.

The third-party verification testing for Lloyd's device is going to cost over $100,000. When all has been said and done, he figures it will have cost many times more than that to run the gadget through the bureaucracy, toward manufacture and the shelf at your local NAPA dealer.

But Lloyd is an optimist with a lot of personal energy and passion— the kind of passion it takes to fire investors' imaginations. And he's already got the champagne on ice.

PLASMA BURNING

No, it's not the liquid part of blood we're talking about. It's that last definition of *plasma*, the one way down at the bottom of the list in the dictionary: "a highly ionized gas containing approximately equal numbers of positive ions and electrons."

A *plasma converter* is one of the black box inventions that made its way to Jeff Voorhis's desk and that he is now particularly enthusiastic about. If he can help this invention see its way to commercial success, it could revolutionize the way we energize the world.

A plasma converter is, essentially, a furnace, but a hellishly hot one at 30,000°—yes, that's *30,000°* F. The technology that produces this exceptional heat, combined with the technology that allows absolutely zero oxygen to filter into the machine during combustion, produces a high-energy state in which matter is *disassociated*. Disassociation in this context means that the matter is torn apart or broken down into its elemental forms—molecules of mainly hydrogen and carbon monoxide. These gases are captured and used in direct heating or cooling applications, to desalinate water, or to run turbines that generate electricity.

What sort of "matter" are we talking about? What is the best feedstock for a plasma converter? Stripped down to its elemental nature, the answer is: garbage. Used tires. Car "graveyards" with their old dashboards and seat upholstery rich in foam and plastics are the pure hydrocarbons the converter is hungry for. Sludge is a fine feedstock. Indeed, Jeff can foresee plasma converters seated next to municipal dumps where the landfills would be mined for a reliable stream of quality fuel for the power-generating unit. The machine can even be fed with animal waste—chicken poop, for instance, an unavoidable by-product of poultry farming—without the need for it to be processed by a digester, which is the common practice now in the mitigation and processing of methane gases for energy.

Now, because different sorts of feedstocks have different Btu values, the makers of plasma converters sometimes suggest that the various feedstocks be processed first through a shredder to achieve a level of consistency. But, in practice, that's not strictly necessary. All sorts of waste—solids, liquids, and gases—can be burned and broken down in a plasma converter, either separately or simultaneously through their own dedicated ports.

The waste materials that can be used, as a matter of fact, include hazardous wastes, such as anthrax or pharmaceutical residues that are currently very expensive to dispose of. That's because when a feedstock is delivered into the plasma converter, only molecules come out.

These molecules take two forms. One is the gases we mentioned, primarily carbon monoxide and hydrogen, or *syngas*, which are used to generate energy. The other is a vitrified, obsidian-like "stone" that is nontoxic and nonleaching. The stone itself is a valuable by-product, useful in road construction or as a kitchen countertop material.

The attribute of plasma converters that you'll likely be most satisfied about is that their technology is emerging as a pristinely clean one. There are no emissions from plasma converters—none, zero, *zilch*. And, when a plasma converter is fed a carbonaceous waste stream (meaning anything from old tires to agricultural residues), it generates the electricity to run itself.

Still not amazed? Think about this: If plasma converters become a conventional part of our renewable energy mix, they will entirely change our concept of waste. There will no longer *be* waste. There will be only feedstock for clean energy. It's a cultural shift that's hard to imagine, accustomed as we have become to our bright green recycling bins. It's *Star Trek*. It's *The Jetsons*. And it's happening now.

Don't get rid of the recycling bin just yet, though. While the technology for plasma converters is, as Jeff puts it, "already there," the plasma people still have a few hurdles to clear. The main one is that a proof-of-concept unit has to be built, and, when it is, it will have to do two things.

1. It will have to prove itself attractive to investors by way of a one-to-four ratio; that is, for example, for every four old tires that are fed into the converter, one of the tires is all that is used to make the electricity that operates the converter while the other three are spent to make electricity that can be sold to the grid.
2. It will have to prove to regulators that the technology can sustain the outcome of the inventors' experiments. When the regulators can sign off, then the plasma converter can move forward to general commercialization.

But which comes first? The construction of the proof-of-concept unit to prove to investors and regulators that the technology is all that its inventors claim it to be? Or the money to build the concept unit that will furnish the proof?

The plasma converter is another of the technologies on the slow track due to slow, low funding for emerging clean technologies.

OUT OF AFRICA

The Seri people of Sonora, in Mexico, roast the stems of the *Jatropha cuneata* plant, split them and soak them, and weave them into artisanal baskets. The species *Jatropha integerrima* is used as an ornamental in the tropics. *Jatropha podagrica* produces a vibrant red dye, and its extracts can be used in the process of tanning leather. The most useful jatropha of all, however, might well be *Jatropha curcas*.

Jatropha curcas, like all of the approximately 175 plants, shrubs, and trees in the genus *Jatropha*, is native to Central American rain forests. It made its way to Africa via Portuguese traders as a valuable hedge plant and now grows wild there, even in the most inhospitable circumstances. Inured to pests, it grows in wastelands, with little water in poor desert soil. Tough, short, and woody, its non-edible seeds have been arduously shelled by hand for centuries to get to their oil, which is used to make soap and is ideal, as well, for candle making since it will burn without smoking. Nutcrackers have now generally replaced the tradition of hand shelling, and clean-burning oil has proved to be just one of the plant's many good points.

You've likely read about the near-miraculous properties scientists have uncovered in many plants native to rain forests, and doubtless already added the preservation of such plants to your long list of reasons why the rain forests must be protected. *Jatropha curcas* is one of those plants. It does indeed have near-miraculous properties.

When the jatropha seeds are pressed, they yield up to 40 percent of their mass in the prized oil. Extracts are already known to be effective in relieving constipation, and scientists at Purdue University in the United States are studying the plant's pharmaceutical potential in treating cancers and HIV/AIDS. After the seeds have been pressed, the cake that remains has many uses: as an animal feed; as fuel for burners to produce electricity; and, because it is rich in nitrogen, phosphorus, and potassium, as a natural fertilizer. The plant's sap is an excellent dressing for wounds, and its leaves can be boiled to make treatments for malaria and fever, or used as food for silkworms.

Every part of the jatropha plant, in short, contains some property that is beneficial to humankind, though perhaps its brightest potential is as a biofuel. Acre for acre, jatropha produces four times as much energy as soybeans, another common biofuel crop, and ten times as much as corn. In tests of various vegetable oils for use in the making of biodiesel, including palm, rapeseed, cottonseed, and sunflower, the lowest emissions were obtained with jatropha. Significantly, because jatropha is inedible, it doesn't compete with its fuel uses for use as a food.

Though the plant has not yet been completely domesticated, there are already large jatropha nursery operations sponsored by women's self-help groups, under a microcredit system, to provide jobs, ease poverty in Africa's rural areas, and begin to create an industry that has the possibility to transform the African economy. Commercial-scale jatropha plantations and the biofuel refineries that accompany them could bring jobs to the continent's rural areas. Also, because most African countries are now oil-dependent, a source of domestic oil would reduce foreign exchange expenditures and associated geopolitical problems. The money saved could be used for future development, which would, in turn, create the need for more energy and help the jatropha biofuel industry continue to expand. It's simply another manifestation of the economic energy cycle: Expending energy intelligently creates the need for more energy, and investing in jatropha would seem to be quite an intelligent choice.

DI Oils, a U.K.-based bioenergy company, has pioneered the cultivation of domesticated strains of jatropha. It has almost half a million acres of jatropha plantations greening with the elite seedlings developed in its horticultural labs. In June 2007 DI Oils announced that it had entered into partnership with BP to produce jatropha-based biofuel. DI Oils is combining its botanical expertise and existing jatropha plantations with BP's worldwide distribution and marketing power in a $160 million investment. As part of the venture, plans are to plant an additional 2 and a half million acres of jatropha in Africa over the next four years, and three-quarter of a million acres every year after that. The biofuel that results is projected to be so abundant that it will meet not only the needs of the local communities; Africa is expected to start exporting biofuel to Europe, where domestic feedstocks are unlikely to be sufficient to support the mandated supply needs for biofuel—11 million tons a year, starting in 2010.

The economic implications of this deal are staggering. Creating an industry at this scale, which produces a product so integral to all world economies as fuel, could transform the continent of Africa into the Saudi Arabia of biofuels.

But wait. Standing at the threshold of a new energy age means that we get to make new rules for new times. We get to start fresh. As Michael says, "We get to do things a little differently this time." Armed with scientific data and the knowledge that comes from experience, we can make better choices for ourselves than perhaps we've made in the past.

Jatropha thrives in arid conditions and poor soil, actually nourishing and improving the soil it grows in. But it grows better in conditions where there is a little more rainfall. This has led some jatropha farmers to burn tracts of rain forest in order to plant their rows in moister, more hospitable conditions.

Alas, these more hospitable conditions are short-lived. Deforestation causes a reduction in rainfall, and the jatropha planted in the former rain forest land eventually requires irrigation. That leaves us with a choice: Do we want to cultivate deserts, or create them? In our opinion, fuel for rain forest is not an acceptable trade-off.

BP and DI Oils have pledged that their jatropha operations will not "compete with food crops for good agricultural land *or adversely impact the rain forest.*"

The italics are ours, to stress a point. With this pledge, we have real hope of growing an industry that will fundamentally change the lives of millions of people for the better, and without further distress to our small blue planet. Of working wisely, in complement with nature rather than in competition with it. Of getting it right this time.

9

THE GLOBAL LOW-CARBON ECONOMY

IN JULY 2005 GORDON BROWN, THEN BRITAIN'S CHANCELLOR OF THE Exchequer, announced that he had asked Sir Nicholas Stern, a former chief economist of the World Bank, to produce a major analysis of the economics of climate change. Brown understood the overarching need to address the world's energy challenge. What he wanted was a comprehensive understanding of the costs involved. Brown saw an unavoidable problem—the growing need to generate more energy coupled with the environmental imperative to do it using clean resources—and he wanted to have the facts at his fingertips. Rather like the rest of us balancing our checkbooks before we pay the monthly bills, he wanted to know where the accounts stood.

Sir Nicholas's report, a 700-page document known as the Stern Review on the Economics of Climate Change, was released in October 2006. While using the strongest possible language to communicate the urgent nature of the energy problem we face as a planet, the report is in many ways a heartening one. We are quite capable, according to Sir Nicholas's calculations, of resolving our energy challenge. And not only that; if we are proactive about it, we have the chance to do it in ways that aren't just reactions to a crisis, but that are actually good for business. "For every £1 invested," Sir Nicholas told the British people, "we can save £5, or possibly more, by acting now."

Conversely, should we shirk our duty and delay, we can expect the global economy to shrink by a minimum of 20 percent.

In the last chapter we sought to give you a taste of the scope of innovation that is happening in response to our energy challenge. Now we want to give you a sense of the scope of worldwide response to the call for a low-carbon economy. We've so far focused, although not exclusively, on the problem and its progress in the context of the developed world, especially North America and the European Union. We turn now to how other countries are addressing our global energy situation. The desire to foster a cooperative spirit in tackling the predicament is more widespread than you have been led to believe—as are the consequences of not moving quickly to get ourselves out of this jam. Just as significant is understanding Sir Nicholas's economic projection. We want to talk about three very different places and the ways in which these countries have internalized the positive economic benefits of *acting now*.

SMALL ISLAND STATES

The United States currently produces 5,956.98 million metric tons of carbon dioxide annually, while all of Europe is responsible for 4,674.75 million metric tons, and China, 5,322.69 million metric tons. These nations, however, are not the ones dealing most immediately or directly with the consequences of carbon pollution. It is the small and vulnerable island nations that are experiencing the first wave of climate change devastation in the form of more hurricanes, typhoons, and cyclones.

Grenada, a nation in the southeastern Caribbean, for instance, once considered immune to hurricanes, has been hit hard twice in the last seven years—and has incurred damage as a result estimated to be in the neighborhood of 200 percent of the country's gross domestic product (GDP). Some countries have already launched programs that will allow them to adapt to these changing climate conditions. For example, in the Maldives, a group of atolls in the Indian Ocean, people are in the process of building fourteen "safe islands." Safe islands incorporate specially constructed shelters that can withstand tsunamis and wind speeds 40 percent higher than traditionally specified for the nation's coastal areas, with plans and drills to consolidate the evacuation of their populations in the event of a natural disaster. These safe islands are being developed at considerable national expense.

The Alliance of Small Island States has requested that the United Nations found an adaptation fund that will assist them in protecting the lives and the livelihoods that stand to be lost in further weather-related catastrophic events caused by excess carbon dioxide in the atmosphere.

Industrialized nations have got to be participants in this effort. They can do so in the long term by cutting their own carbon emissions, of course, and in the short term by supporting such an adaptation fund. To do less would be a gross failure of responsibility—rather more severe than opening up your car window and tossing a handful of litter into the wind and expecting the local community to clean up after you, but you get the point. To do as the small nations ask will prevent not only their potential losses but both the expense and the heartache of having to send relief workers to clean up after a preventable tragedy.

CHINA

With a rate of economic expansion that is putting the country in the unenviable position of racing with the United States to be the world's biggest emitter of greenhouse gases, China is sometimes viewed as a reluctant participant in the global energy challenge, as too willing to put profit above pollution mitigation. But while the United States was struggling to pass renewable energy legislation in Congress in December 2007, a comprehensive renewable energy package had already been law in China for two years.

Under the law, the state officially encourages the construction of renewable energy facilities through both direct funding and tax incentives, and the country's grid is obligated to purchase all of the electricity generated by these renewable sources. This law establishes the framework for provincial planning agencies to develop specific ways to implement the law locally by providing a target—15 percent of all of the country's electricity is to be generated through renewable sources by 2020—and specific penalties for noncompliance. As a consequence, China's renewable energy program has grown at an average annual rate of 25 percent over the last few years. How does that translate in terms of dollars and cents? In 2005, when a total of $38 billion was invested in renewable energy worldwide, China's commitment of $6 billion topped the spending list.

After several decades of economic growth at seemingly all costs, China has pointedly expressed its interest to "try its best to address the common challenges faced by the international community," in the words of one Foreign Ministry official.

Why? And why now?

Part of the reason is undoubtedly inspired by patriotism. As the world's third largest economy, Beijing has a great desire to be seen as a responsible world power. That's a goal that's undermined now, in large part, by oil. China, once oil independent, these days must import a majority of its oil supply to meet its energy needs, and it does so from Sudan. With the terrible humanitarian crisis in Sudan's Darfur region, China is under pressure from other countries to join in imposing sanctions on its key supplier of oil. This is just one example of how oil dependence limits the strategic options of countries all over the world, and it is a situation that cannot be remedied until investment is made in growing the renewable power industry.

Another factor in China's embrace of green energy is that it is home to twenty of the thirty most polluted cities in the world. There is a rising tide of public anger about the country's gray skies. Estimates are that its choking smog causes an approximate, appalling 400,000 premature deaths every year. In preparation to host the 2008 Olympics, the government in Beijing did more than restore and beautify cultural landmarks such as the five famous pavilions in Jingshan Park and the Fire God Taoist Temple—it conducted experiments that included banning up to a million cars a day from driving on its roads in order to improve the city's air quality.

Perhaps, however, as with all countries, the most practical reason is the most pertinent. If China aims to attain energy security by maintaining a high level of energy self-sufficiency—and it does—it's got a more immediate challenge than many other nations. Seventy percent of China's energy needs are now met with coal. While most of the nations of the world can rest assured that they have a 200- to 300-year supply of coal on hand, China's coal mines are expected to be mined out within only 80 years.

Several barriers stand in the way of a more rapid deployment of renewable energy in China. One is a lack of independent development technology. China now imports both the technology and the equipment used in its renewable power programs. Lowering the cost of renewables production and consumption through research, local manufacture, and the sharing of technology—especially clean coal technologies—is central to overcoming the problem.

Another core difficulty is the lack of specialized workers. Implementing even a portion of China's wind power capacity alone, for instance, would require skilled workers by the tens of thousands, but in the country's nearly one thousand institutions of higher education, only one provides for a four-year program in wind energy.

Still, Beijing recognizes that an economy built around clean, domestic energy sources is far more secure than one built around a dirty fuel whose supplies are fast running out. It is moving, however gradually, to adapt to the coming new energy age, and it needs to be engaged in the international solution with respect for both its determination to pull its people out of poverty by continuing to grow a vibrant economy and for the giant steps it has already taken toward making that economy a low-carbon one.

ABU DHABI

Abu Dhabi is a place-name associated in the world's conscious with oil wealth so extravagant it seems almost as dreamlike as one of Dubai's newest hotels, built to resemble the billowed sail of a ship. In that context, it might be surprising to learn of the extent of its commitment to renewable energy technologies. Abu Dhabi's "Masdar" alternative energy initiative is possibly the most comprehensive response to the global energy crisis anywhere in the world.

Al Masdar comes from the Arabic for "source." Under the Masdar initiative, the government has made available a four-square-mile campus and a $150 million clean technology fund to find new sources of power for the coming new energy age. On the campus, a research headquarters is being built in which scientists and engineers from the United Arab Emirates (UAE) will carry out research in conjunction with the institute's postgraduate education program and in collaboration with its research partners—distinguished institutions like the United Kingdom's Imperial College London, RWTH Aachen University in Germany, Canada's University of Waterloo, Columbia University in the United States, the German Aerospace Center, and the Tokyo Institute of Technology in Japan. Through the initiative's special economic zone, the money in the clean tech fund will be co-invested in renewables with private sector partners that include some of the world's largest energy companies: BP, Shell, Occidental Petroleum, General Electric, Mitsubishi, and Rolls-Royce. The Masdar's "innovation center" will support the demonstration, commercialization, and adoption of carbon mitigation and sustainable energy technologies.

To draw the world's attention to its ambitious program, and to foster the cooperative exchange with elite educational institutions and large industries that such a program requires to succeed, in January 2008 Abu Dhabi was host to the World Future Energy Summit. The summit drew delegates and exhibitors for a three-day event that featured the Prince of Wales as a keynote speaker, and mixed political debate with talk of technical solutions and investment opportunities in service of determining the global strategies that will take us into the post–fossil fuel age—and all of this is taking place in the heart of the world's fossil fuel capital.

It's not business as usual in Abu Dhabi, and the reason why is simple. "UAE is a very successful country but realizes that its wealth is based on a finite resource," explains Peter Evans of Imperial College's Energy Future lab. "The Masdar project takes a long-term view for a sustainable economy."

In Abu Dhabi, they are fostering advances in renewable technologies in a unique and multifaceted forum that will allow them to retain, and even to grow, their share of the global energy market as that market adapts itself to new energy realities.

India is another emerging world economy where growth in consumption has caused an alarming increase in its dependence on fossil fuels. The country's Ministry of New and Renewable Energy operates a revolving fund for the development and deployment of renewable technologies that is helping India to reach its goal of 10 percent additional installed renewable capacity by 2012. In Saudi Arabia, another obviously oil-rich country, it has been determined that it would be highly uneconomical to expand the existing electrical grid into sparsely populated regions. Two prime areas have been identified for the installation of wind power, the blustery Arabian Gulf and Red Sea coastlines, and testing to develop these sites is under way. In New Zealand, hydroelectric generation and geothermal power already provide so much of the country's energy they are referred to as "traditional" renewable energy sources, and the country is now able to explore a new range of "nontraditional" renewables.

What is happening, in short, all over the world, is exactly what Yvo de Boer, executive secretary of the United Nations Climate Change Convention, called on to happen at the 2007 conference in Bali: Countries are getting clear on the instruments they have at their disposal to act on mitigating climate change and adapting to the new energy reality. What resources are realistically in hand? What market mechanisms can be put into place to encourage compliance? What government support structures are necessary to create the market certainty that will foster investment in renewables from the private sector—a sector, by the way, that is projected to supply 86 percent of all funding for renewable research and deployment?

Most important, we can see that countries are approaching carbon mitigation and renewable energy technologies not as separate from good business practice but as part of it.

Total investments in new physical energy assets are expected to triple between the years 2000 and 2030. How do we go about developing an investment strategy for this massive, global undertaking that will assure us of the most value for our dollars? How do we reorient ourselves, restructure old ways of thinking about energy investment that have been in place for generations? How do we, in the words of Mr. de Boer, alter "the course of an investment supertanker"? Sir Nicholas puts the costs of securing our environmental and energy future at 1 percent of global GDP per year—a modest sum in light of the global challenge it is meant to resolve. How do we generate and allocate the funds to do it?

It will take intelligent financial planning within the framework of a cooperative global strategy—nothing short of the urgent investment of resources and will that surrounded the Manhattan Project or launched the Apollo program, efforts to which the global energy challenge have often been compared. We like Mr. de Boer's metaphor best of all. Securing our energy future will be like "embarking on a Star Trek expedition, making public and especially private money go where it has never gone before."

10

THE GREEN STANDARD

BEST PRACTICES

WE'VE COME TO THE PLACE IN THIS BOOK WHERE WE PRESENT OUR CALL FOR action, the very serious business of outlining the steps we must take now to address our energy challenge. If we don't take them soon and with confidence, we may not have the time to react to the problems in ways that preserve our earth and the standard of living we enjoy on it. This, of course, has been the whole intent of the book—to cut through the plethora of information and misinformation, facts and biases surrounding the urgent energy issues we face as a planet and provide a practical guide to resolving them. Before we do, we want to review what we think of as the key points that help to bring our global predicament into human perspective, into a scale that makes the actions we need to take manageable.

First of all, let's talk about what we mean by "standard of living." We're not referring at all to the shiny new cars or fast computers or even the comfortable, climate-controlled homes many of us expect as a matter of course here in the developed world. We're talking about something much more basic and much broader. We're talking about air that is fit to breathe, planetary temperatures that sustain our habitats, and weather patterns that don't destroy them. These are all striking, classic environmental issues. However jarring or ironic you may find it, none of them can be solved with environmental actions alone. These are all matters of energy—how we make it and how we use it—and answers to energy problems will come only from changing the ways in which we conduct business.

What do we mean by "changing the ways in which we conduct business"? Well, on a personal level, it means being energy proactive in your home and workplace—changing the ways you go about the daily business of lighting and heating, landscaping and driving to the office. There are any number of sources for this sort of advice—Web sites on how to conserve energy, magazine articles on ways to make your home more energy efficient, newspaper articles that speak seasonally on how to celebrate a green Christmas or plant a water-conserving garden.

Energy conservation, if you'll recall, is an integral one-third of the formula we believe we all need to follow to achieve energy security. So don't think for a minute that we're giving it short shrift here by not including in our list of best practices, all of those many tasks you can accomplish in your home or business to reduce the amount of energy you use immediately. On the contrary, please change the light bulbs in your living room if you haven't already switched to compact fluorescent bulbs; if every home in America changed just one of its old-fashioned light bulbs for an energy-conserving one, that would save in just one year enough energy to light more than 3 million homes, save more than $600 million in annual energy costs, and save more carbon dioxide from entering the atmosphere than had 800,000 cars been taken off the road.

One little light can do all that!

You see how important the bulb in your reading lamp can be.

And please consider, as you remodel or build your home or office, renovating and retrofitting green. Energy Star appliances, passive solar water heating systems, improved insulation and weatherstripping on doors and windows will save energy and money, even if sometimes these monetary savings are realized only in the long term. Thoughtful upgrading of the way they use energy, in fact, saved Wal-Mart the expense of an estimated 600,000 kilowatt-hours a year when the company simply enclosed the open-air refrigerated sections in its retail stores.

Most of us aren't in a position to directly affect energy consumption policy at the world's largest retailer, but we are in a postion to either suggest or directly implement kilowatt-saving measures in our own workplaces. The 10 or 15 kilowatts you save a month by unplugging your computer before you go home in the evening builds on the 15 or 20 kilowatts your neighbor saves by retrofitting his signage to use less electricity builds on the 20 or 25 kilowatts his neighbor saves by turning her thermostat down two degrees. Pretty soon you're talking about some real kilowatts!

You needn't stop with your home or business, either. Ask if your child's school is green. If it isn't, try logging on to www.greenschool. com for suggestions on how to make it so. Take the information you've gathered to your child's principal or another school administrator, or form a parent group that can help you devise and act on a plan to make your child's place of education a green and healthy one. Partner with a local company that is likely to be just as interested in helping your school conserve energy as you are—if for no other reason than the positive public relations that it will generate—and that can supply funds and/or materials to make your green school happen.

There are two key points to keep in mind as you explore and implement the ways in which you can make your home, business, or school more energy efficient. Our good friend, Rob Watson is the "Father of LEED," the gold standard of green building rating systems and one of the leading experts in the international high-performance green building movement. His illustration of these points is potent and succinct.

You don't have to give up quality in your quest for green

Let's go back to that bulb in your reading lamp. Most people prefer warm light in their homes and the new compact fluorescent bulbs can leave some feeling cold. Rob tells us that this has to do with kelvins—a unit used to measure temperature. To enjoy warm light from our new compact fluorescent bulbs we should look for ones that are labeled 3000k or less, with 2700k being optimum.

Bulbs with lower kelvin numbers often cost a bit more than bulbs with a higher kelvin rate, and that brings us to the second key point.

Conservation isn't an expensive proposition

At least, it doesn't have to be if you look beyond the price of something to its true cost. Most of us have gown used to paying around $2 for a light bulb. Compact fluorescent bulbs are a more expensive purchase—let's say $6 for a 2700k one for the sake of our discussion; compact fluorescent bulbs that have higher kelvin numbers are less costly—let's say $4 to keep the math simple. It might be natural for us, trained as we are to look at price, to opt for the $4 bulb. When we get that $4 bulb back home, however, the light it sheds isn't as comforting a one as we're used to reading by, and we begin to believe that harsh light is a price we have to pay for energy efficiency.

This isn't true. Not if we can retrain ourselves to consider cost instead of price. Old-fashioned incandescent light bulbs generally have a lifespan of 1,500 hours; compact fluorescent bulbs have a lifespan of 10,000 hours. If you follow the math on that little fact, you'll see that it's possible to spend $14 to get 10,000 hours of illumination from an incandescent bulb, or $4–6 to get 10,000 hours from a compact fluorescent one. This savings alone makes the compact fluorescent bulb a no-brainer, but compact fluorescent bulbs also require fewer watts to provide the same light intensity. When you factor in the savings on electricity—a savings that, depending on things like the number of lamps you have in your home, can translate into a 75 percent reduction on your light bill—you'll see that it's really not an enormous splurge to go ahead and treat yourself to the $6 bulb that will provide soft, comforting warm light.

"It's a matter of short term versus long term, first costs versus life-cycle costs," says Rob—a matter of reorienting ourselves to take in the big picture rather than simply responding with sticker shock when we go to the hardware store to buy our new bulbs.

But, as we've said, no one part of the formula can stand alone, and here's the catch to conservation: We tend to use the energy we conserve by finding ever more ways to consume it. "Energy intensity" is how economists calculate the energy efficiency of a country's economy. It's measured as units of energy used per unit of gross domestic product (GDP) and it's influenced by a variety of factors, including standards of living and weather conditions.

In the United States, for example, the country's energy intensity has dropped by nearly 50 percent since 1975. Through both an awareness of how we use energy and technological improvements, we've slashed that number in half. But because, at the same time, the population has grown, and the economy has grown, and we have come to rely to an ever greater degree on energy-consuming products like microwave ovens and personal computers, energy consumption has risen by 40 percent. Think of this striking fact: In the last sixty years, the average home has grown from 1,000 square feet to 2,500 square feet. Over double the old average size. All of that additional space must be lighted, cooled and heated, vacuumed, carpeted, furnished—and all of those things increase the energy these larger homes require on a daily basis.

So, then, if energy conservation tends to be offset by increases in overall usage, energy conservation can't be the lone or final objective. We have to build our knowledge of renewable energy technologies through research, and we have to build the renewable energy facilities themselves. This part of the formula we propose for a secure energy future is going to take money to make it real. The amount of money that's available to invest in making our secure energy future a reality will be, in the end, determined by the wisdom of the policies set by our governments.

We want our list of best practices to be about empowering—indeed, *emboldening*—individual citizens to take the energy policies of their local, regional, and federal governments very personally. It may seem a curious paradox, but these government policies will be implemented only when enough individuals put their voices together and tell their leaders to do the right thing. This chorus of voices is beginning to sound all over the planet: from people in China who are fed up with choking on air, to people in the United States who have had it with carmakers who won't give them vehicles that get the gas mileage cars are perfectly capable of achieving with nothing more than a little bit of retooling and a little bit of will.

The results of the people raising their voices and declaring that they want our energy future to be given top priority—that, indeed, they want it to *begin*—are being seen in elections all over the globe. It's the people who are telling their leaders to get on the renewable bandwagon or get off the parade route. Our planet and our economies are under startling pressure right now. Only by keeping up the same sort of intense pressure on our elected officials will we see put into place the policies that will get us out of this jam.

Hand in hand with making it a requirement that, in return for the privilege of leading us, our leaders must sign on sincerely and proactively to solving our energy problems, however, is the understanding that the demands coming from the people must be reasonable ones.

Neither half-thought-out ideas nor half-baked schemes are going to provide the timely and economy-friendly solutions we need.

We've striven, in these pages, to provide an overview of the abundance of technological solutions that are available to apply to our energy challenge. In assessing how they will be most useful to us, we have to stress again that three important factors must be taken into consideration.

1. **No one renewable energy source will be appropriate to every location**. Geographical features and weather patterns, as well as man-made assets such as existing dams and reservoirs, will all be a part of the thought process in determining which renewable source or sources best fit the needs of any specific community. Towns, cities, states, provinces, and nations all need to assess for themselves the assets they have on hand and how to productively incorporate them into their own unique energy plan. Wind turbines might be just the thing for Saudi Arabia's Red Sea coastline but they may be less than feasible for downtown Atlanta, Georgia, to give a very broad example of what we're talking about. But the rooftops of Atlanta may provide just the surface area necessary for a successful and grand photovoltaic installation while the sun-baked desserts of the Middle East could prove to be ill-considered as a site for solar electricity generation because dust storms may impair the efficacy of the equipment.

2. **No one renewable energy source can likely do it all alone.** Wind and solar, geothermal and biomass, or sugar ethanol and cellulosic ethanol deployed in tandem have a better chance of meeting our need for reliable and high-quality power and fuel than they do when they stand alone. Local renewable programs—which are, as we've discussed, where the heart of our renewable future lies—will serve their populations best when they employ several technologies that complement one another.

3. **Timing is everything**. Step back and take a breath. The sense of panic that's fostered by the dire warnings of our

planet's imminent peril could actually do more harm than good if they cause us to rush our leaders headlong into ill- or only partially-conceived plans for action. In Bali, in December 2007, the United Nations Climate Change Convention set, as one of its goals, a two-year time limit to formulate a global plan for action. Time is, assuredly, ticking. But this time constraint seems to us to be even more of a reason to use what time we have well, for setting a reasonable course of action.

Think of it this way: When a man experiences chest pains, there may well be a natural tendency to panic. But clearer heads that take a second or two to think through what is happening will likely remember that the first thing to do is to give him an aspirin—an aspirin taken at the first sign of a heart attack can prevent further damage to the muscle by thinning the blood. We need to take that second or two, clear our heads, take an aspirin.

We need to reorganize old ways of thinking and get clear on what the new ways of approaching our energy problem will make possible. We need to have a fundamental grasp of the technologies available to us so that we can deploy them reasonably: efficiently, reliably, and cost effectively. We have to decide how we are going to fund the deployment in an equitable and economically friendly manner—and we don't have to reinvent the wheel to settle on our methods of funding either. We simply have to reach back to our history to find solutions that worked and adapt them to our current circumstances. A cap-and-trade program worked beautifully to solve a grave sulfur emission problem in the United States in the 1990s. If we take a page from the Rural Electrification Administration, we can find ways to fund today the sort of visionary advances that Roosevelt helped make possible in the 1930s.

Only when we have taken these thoughtful, preparatory steps can we confidently ask the government to set the policies that will provide the framework for the new energy age. This is because when we make these policies, we have to be prepared to stick to them. Only when the renewable market is certain that governments are committed for the long term to the new energy age will there be investment in development and deployment of the technologies on the scale we need to meet our energy challenge. It's a simple market equation: Renewable energy has to become as profitable for its producer as fossil fuel energy is now, or the producer won't want to make it, and you won't want to buy it. Renewable energy technologies, like other energy technologies that came before it to evolve the way we make and use energy, require policy support to become profitable.

Here, then, are some well-thought-out guidelines for government policies that will have the power to take us out of the old age and into the new.

CARS

We've already said it once in these pages, but it's so central to responsible energy management that it bears repeating: The single smartest move we can make to ease our transition to the new energy age is to raise the Corporate Average Fuel Economy (CAFE) standards. In late 2007, the United States took a small step in that direction.

The CAFE standards, basically, establish the minimum number of miles a vehicle must be able to travel on one gallon of gas. In the United States, that standard was a mere 27.5 miles for cars and 22.2 for light trucks. These were minimum standards that, in spite of technological advancements and increasing environmental outcry, hadn't been raised since 1975. In December 2007, the United States Congress passed a still slim and timid energy bill. The bill's most dramatic component was the largest change in the fuel efficiency standards in over thirty years. The combined average of passenger car and light truck fuel efficiency for new vehicles is now mandated to reach 35 miles per gallon by the year 2020.

The importance of raising the CAFE standards is fairly obvious: The farther a car can travel on a gallon of gas, the less gas has to go into its tank. But the main reason we had to do it first, before we even thought about doing anything else, was because we *could*. Twenty-five miles per gallon, however, is nowhere near as far as carmakers are capable of making engines go and the truth is that thirty-five miles per gallon is not itself an ambitious standard. The technology is available to squeeze fifty miles or more out of a gallon of gas—technology that would, in essence, reduce both carbon emissions and your monthly gasoline bill. Cars that make the most of every drop of fossil fuel we put into them can bridge our transition from the gasoline era into the new era of biofuels. We can't stop at thirty-five miles per gallon. We need to keep pushing the technology, incorporating it faster into the cars we drive, raising our fuel efficiency standards into the realm of what is really possible. It is unconscionable that lawmakers stymie this most fundamental of actions—one that we are able to take immediately to benefit both our natural and our economic environments.

Let's not, however, put all of the blame for this lack of fuel efficiency on the carmakers. As a group they have lobbied hard against raising the CAFE standards, true. But they are also an industry with stockholders to satisfy and jobs to protect. Raising the CAFE standards should be done in conjunction with a federal subsidy plan to help carmakers retool their machinery and generally gear up, so to speak, to produce the more modern cars we need while we cross the bridge to biofuels.

The next step in transitioning the transportation sector to the biofuel era is to address our fuel delivery infrastructure. How can we most efficiently get biodiesel and, particularly, ethanol, to the consumer reliably and on a mass scale? This is not a call for lawmakers merely to increase biofuel production goals; they need to take a hard look at laying down an ethanol pipeline system, a network of pipes made of material that will not corrode when exposed to the water that ethanol attracts. And it's a call as well for lawmakers to begin to require that fueling stations be equipped with Flex Fuel pumps that will allow consumers to choose between a mixture of gasoline and ethanol, from 10 to 85 percent, or pure ethanol alone. While at least for a few years, there might not be many cars on the road that can travel on 100 percent ethanol, the only way a transition can be made to a pure ethanol era is, of course, by taking the first steps toward the goal—and drivers can't fill up their cars with biofuel until they can access a biofuel at the local gas station. Pipelines and pumps that can deliver and dispense the new fuels are critical. To put it plainly, we can produce all of the biofuel we want, but if the consumer has no way to buy it conveniently, it's a product that will have a very hard time finding its market.

What is the strategy for the construction of such a pipeline system and for the installation of modern Flex Fuel pumps? Sadly, few studies have been completed from which a coherent strategy can be developed. Such transition studies should be commissioned, and with all due urgency, by federal governments. A comprehensive calculation of the costs involved in these national projects must necessarily include incentives for pipeline and fueling station owners to ease their compliance with the new delivery and distribution regulations that will follow from the studies—after all, these are business owners who have been playing the gasoline game by a certain set of rules. When we change the rules, it's not only fair to them but advantageous to us to make sure the seasoned players can stay in the game.

At the same time that strategies are being developed to update pipes and pumps, more funds need to be directed to research in cellulosic ethanol. The first cellulosic ethanol distilleries are up and running, but developing the technology to the scale needed to meet global fuel demand will take some real money. Compare the $1.4 billion that went to subsidize the oil and gas sectors in the 2006 U.S. federal budget with the $91 million allotted for research in cellulosic technology for an idea of how our fuel priorities are misplaced and how our environment and economies are compromised.

The development of cellulosic ethanol should not be limited to the production of the fuel alone. The scope of research projects must include broader applications for the technology. What by-products will come from converting wood waste or switchgrass or lawn clippings or grape stems into fuel? Electricity from burning plant waste that can be used to drive the machinery in the production of ethanol as well as extra that can be fed into the grid? Animal feed? What biochemical compounds can be extracted and used as building blocks in the manufacture of other products—plastics or pharmaceuticals, candles or crayons or eyeglass frames?

But we should not—and, really, *can*not—wait until cellulosic technology is developed to scale, for a mass market, to begin our transition to biofueled transport. As a bridge to a time when there is general availability of cellulosic ethanol, we need to begin using ethanol made from other feedstocks. Corn ethanol is impractical for use in even such a temporary transition period—the gains of using corn-based biofuel are outweighed by the drawbacks of the fossil fuels necessary to produce the bioproduct, and that defeats the whole purpose.

But as a country's farmers and ethanol distilleries transition to cellulosic feedstocks and production capabilities, ethanol can be imported to flow through the new pipelines and accustom drivers to the new fuel. Sugar ethanol from Brazil, a producer with a ready, steady supply, is the obvious choice. In order to do this in the United States, the prohibitive tariff on Brazilian sugar ethanol will need to be removed.

This tariff—54 cents a gallon as of this writing—was put into place in order to protect American ethanol producers and the domestic market for their product. This protection might make sense if America was in a position to provide its citizens with a steady supply of truly environmentally sound and cost-effective ethanol product. But it's not. Not yet. And the purpose of the tariff has since become a bit convoluted, confused with a natural and strong desire to promote domestic business operations and the similarly natural and strong desire for energy independence.

Look at it this way: America right now imports twelve and a half million barrels of oil a day. Much of this oil carries with it limitations on geopolitical options that energy dependency can impose. But somehow it has become part of the conventional wisdom that importing ethanol is a bad thing—that from the outset of what is in America an infant industry, all ethanol supplies should be domestic, no matter the quality of the product or the current capacity of the industry. Holding ethanol to a higher importation standard than oil even while the country must continue to import polluting petroleum is like, well—it's like shooting yourself in the foot.

The final piece of the transportation sector puzzle is to convert the world's fleet of vehicles to Flex Fuel models. This will naturally be a slow process. We have a friend who is still holding on to the little red truck he bought back when he was in college; it was used when he bought it during the Reagan administration. Now, while he drives the truck these days only on weekends—and only when his wife isn't going to ride along—and while we realize that not everyone has such a sentimental attachment to his first set of wheels, the point is that it will likely take many years until Flex Fuel vehicles become common. There are, however, many ways in which federal, state, and even local governments can speed up the process. Here again, we can all take a few tips from Brazil:

- Convert government fleets to Flex Fuel vehicles.
- Offer tax incentives for corporate cars, taxi fleets, and to individual consumers who purchase Flex Fuel vehicles.
- Subsidize the retooling and other modifications that car manufacturers will have to make to their production lines in order to make it profitable for them to adapt to a new norm in transportation.

Getting ourselves off gas won't be easy. The process is multifaceted, and all of those facets have to be coordinated so that every part of the transportation sector will be adapting in sync with each of the other parts. While the complexity can seem intimidating, it is not so much more, we think, than any other technology that has required us to construct or adapt an infrastructure and change the way we think about and perform ordinary tasks. The Internet was once a scary place for those of us accustomed to typing letters and sealing sheets of paper in an envelope before we could communicate in writing with another living soul, but now I'll bet even your grandma checks her e-mail when she gets up in the morning.

CARBON

In 1997, the Kyoto Protocol put a price on carbon, establishing it as a tradable commodity. The successor conference to Kyoto, in Bali in 2007, recommended a 25 to 40 percent reduction of carbon emissions by the year 2020 as realistic and reachable for industrialized countries. How do we take advantage of the new commodity—carbon—and make the best use of the market it's generating to reach our carbon reduction goals? As we discussed in chapter 7, our best tools are carbon offsets, green tags, and cap-and-trade programs.

Carbon offsetting is an affordable way for an individual to mitigate the carbon she is responsible for generating when she takes a flight home to visit her favorite aunt or chops down the maple tree in her backyard that is sending roots under her house and compromising its foundation. Unfortunately, carbon offsetting is pretty much a Wild West industry right now, generally unregulated, and that undermines the confidence a consumer wants to have when she's investing even small sums of money in projects she believes will help to make the world a little bit better place. No one wants to just hand over hard-earned dollars to some cowboy and hope that they're going to be spent reasonably. A process of review and accreditation for projects receiving carbon offset funds and a national or even an international registration or licensing system for companies that trade in carbon offsets are called for to inspire consumer confidence and participation in these sorts of programs.

Green tags, we suppose you could say, create a commodity that is actually just the opposite of carbon. By separating the attributes of renewable energy—the electricity, for example, from the environmental benefits of generating the electricity with a green resource—they allow each attribute to be sold separately as well. My power provider can sell me the electricity it purchases from a solar field, and you can buy the property rights to the environmental aspect of that same electricity. Like carbon offsets, green tags are an entirely voluntary program, and the green tag industry is just beginning to become more formalized, with its own set of regulations and regional over-sight systems.

There are a few drawbacks to green tags, and most of them—the potential for double counting being the most problematic—could be solved by putting in place federal policies that standardize the rules under which green tags can be bought and sold, and instituting a national registry of green tags, or renewable energy certificates, ID numbers. The suggestions we make in reference to expanding the oversight of carbon offsets and green tags aren't, of course, for the purpose of idly increasing the administrative costs of these pro-grams—and such management will most assuredly do that to some, we hope lesser, degree. We make the suggestions as a matter of basic con-sumer protection. When money is voluntarily given over to assist in the solution to a problem as momentous as our energy challenge, the only moral thing to do is to make sure that the money ends up going where it is intended to go.

Cap-and-trade programs are the way we'd prefer to see major industrial polluters held responsible for the emissions they create. Taxing carbon emissions directly may seem to be the more straightfor-ward approach, but market systems like cap-and-trade have proven themselves to be not only a more equitable but a much more economy-friendly approach to solving emissions problems.

There are two keys to imbuing cap-and-trade with the substance and vitality it must have to be as useful a tool as we need it to be. The first is that the cap that is set must be *mandatory*. The only way that a cap-and-trade policy—or, really, any policy we design to alleviate our energy crisis—is going to work is to give that policy teeth. The policy has to be attached to penalties for noncompliance, and the penalties must be vigorously enforced.

In January 2007 Massachusetts governor Deval Patrick gave a bit more bite to the Regional Greenhouse Gas Initiative (RGGI), an orga-nization of seven states that aims to reduce carbon emissions in the northeast, when he signed an agreement to begin assessing a penalty for every pound of emissions released in his state. He predicted that the costs of electricity would rise, initially, by about $3 per household per year because of the fines that would now be levied on polluters. But he also predicted that within a few years consumers could be saving up to $70 a year or more due to the conservation efforts the penalties would inspire the polluters to take. And, in the meantime, over $125 million would be realized in penalties—money the state will use to fund the expansion of its renewable program.

The second key may be a bit harder to come by, but it is nonetheless critical, and we need to strive as an international community to attain it. This second key is international agreement on the caps we set.

This does not mean that the caps must be uniform for every nation. It would be unjust to expect that a country producing 2 percent of the world's carbon emissions be held to the same reduction rate as one that produces 25 percent. Nor does it mean that every country should be held to the same timeline for achieving its agreed-on reduction. Richer countries have more assets to put to work to produce results and might well be held to a higher performance standard, and/or expected to assist countries with poorer economies to meet their own goals.

As the caps are set on an international level, what the negotiators have to keep in mind—what we all have to keep in mind—is that our energy crunch is no one country's national problem. It is a global problem. Grenada pays in the currency of hurricanes for carbon emissions that originate in France. Green Bay feels the temperature differential to which the factory in Beijing contributes. International accord on caps will produce two important results. It will curtail the potential for the sort of "leakage"—polluters getting around carbon caps by moving their operations to locations where caps do not exist—that we talked about in chapter 7, and it will underpin the cooperative resolve of the world in tackling and defeating a global crisis.

RENEWABLES

Renewable sources of energy are one of the three backbones of achieving energy security. These are the new technologies we must come to rely on if we are to realize a clean energy future and create in full flower the green industry that is already beginning to enhance and grow our economies. Geothermal plants, solar fields, wind farms, biomass operations, and hydroelectric power stations are the coal plants and oil fields of tomorrow, if you will. So that we can stay abreast of our growing need for energy, we must start to build more and more renewable facilities today. Our governments, particularly our federal ones, must step up to the task.

The best first contribution a federal government can make to its country's energy advancement is to put into place a policy of mandated renewable energy standards (RES), also referred to as renewable portfolio standards (RPS). Such a policy would require that each power provider creates or purchases a certain percentage of its energy from renewable sources and includes this green power in its mix of system power. As of this writing, 15 percent is on the high end of average.

In the United States, two dozen states have already enacted their own RES, though no national policy yet exists. These state programs demonstrate vision and determination that governments at the federal level would do well to emulate, and here's why. Power distributors are often reluctant to contract with green power producers because green energy can, at this time, come with a higher price tag than fossil fuel power. In mandating that the green power producer's product will have a buyer, the policy creates a market certainty that gives confidence to investors. That, in turn, leads to more available funds to build more green power plants. Through building more green power plants—deploying the technology more commonly—the price of green power is gradually reduced. And as the price of green energy comes more and more into line with its fossil fuel counterparts, the need to use fossil fuels is gradually displaced. Renewable energy standards are what kick-start this positive cycle.

What can keep the cycle in constant motion are the expectation of excellence and flexibility. By "expectation of excellence" we mean periodic review of the RES that starts from the premise that it expects a renewable goal will be met, determines that it has been, and raises the goal. That's how we get, slowly but steadily, from 15 percent renewably generated power to 20 percent, and then to 25 percent, and so on.

Flexibility involves consideration for a state or province that might have fewer resources at hand that lend themselves to renewable adaptations than its neighbor does, and targeting that location for additional funding. Or it might mean allowing a state that can more easily meet a set goal to increase of its own accord its renewable standards in excess of the federal mandate—perhaps even rewarding the state for surpassing the goal.

Following close on the heels of adopting an RES policy comes a sustained federal commitment to green energy research and production. This means giving financial support to programs that, for example, are developing the next generation of wind turbines or investigating the environmental feasibility of enhanced geothermal. It means also giving financial support to those green technologies that don't exist yet. Perhaps one way to do this is through policies under which black box reviewers can be conduits in applying for grants or low-interest loans that will see a promising new invention through expensive third-party verification processes, and/or enacting policies that reduce not the verification process itself but much of the red tape associated with moving each promising invention through each successive phase of its life.

The rewards of research like this have the potential to be even more far-reaching when the results are shared with the global community. The vibrant exchange of research information while a project is in progress—that is, formal and informal collaboration—could lead more brains to more breakthroughs more rapidly. The transparent exchange of information about breakthroughs, when they happen, could lead to more rapid deployment in every region of the globe where the particular technology is applicable. And sharing research is just one way—but a crucial way—in which richer countries, which can better afford to make the investment in research, can help turn the tide both literally and figuratively for a country unable to support its own renewables development but desperately in need of renewable technology.

Federal commitment to the production of renewable energy is the third pillar of responsible federal response to our energy challenge. This means a commitment to fund construction of green power plants at some certain and sustained level. It's not only the funds themselves that are important in this sort of commitment, though without doubt the funds are necessary. It's market certainty, again, that is among the primary goals of any funding initiative. Public funding helps to create the favorable business climate that attracts private investors and creates another positive cycle—this one of investment-return-reinvestment—that will enable the growing green industry to flourish.

At the regional, state, and local levels, legislators can be more specific about taking into account the strengths of each renewable resource in regard to each area's particular needs and assets. Here we're talking about programs like Berkeley's solar roof initiative, with its first-in-the-country "Sustainable Energy Financing District." An undertaking like Berkeley's exploits a natural solar asset specific to its area, and can undoubtedly be administered more competently and at less cost at a local level. But we have to wonder how such programs could be enhanced and expanded by a strong federal framework that offers support, and possibly even reward, for communities that demonstrate such brilliant and practical green innovation.

CARBON SEQUESTRATION

As we discussed in the section on clean coal in chapter 2, there are two types of carbon sequestration. The first is terrestrial—meaning the carbon that is stored in the natural life cycle of carbon-based life-forms living and dying. Carbon is taken in by plant life in the oceans, by wetlands, agricultural lands, and forests. It is taken in effectively, for example, by growing trees, and stored there until the tree dies or is cut down. It seems like an elementary process—plant a tree/store carbon—but, as we explained, it's a little more complex than that.

Planting forests on a grand scale, with appropriate species of trees, and then thinning them at regular intervals so they retain their efficacy as carbon sinks is but one facet of land management. Preventing deforestation is another aspect of terrestrial sequestration so vital that, at the two-week climate conference in Bali, one entire day was devoted to the subject. Funds well spent are directed at professionals trained in the science of land management and skilled in its art.

The second form of carbon sequestration involves recapturing released carbon at its source and storing it in cavities within the earth or deep beneath the ocean. This form of sequestration has generated a great deal of press coverage, at least in proportion to its titillation factor. It's a subject that's a bit drier, we admit, than Britney Spears's latest adventure.

The reason people can like to talk about carbon sequestration is that it is a hopeful technology, a prospect that ignites the human desire to right a wrong, clean up a mess, and finally tidy up after a long and decadent party. Conversely, it brings a certain comfort along with a wishful thought: We can just go right on using coal as a fuel because soon we'll know how to use it without polluting, and that means we don't have to do anything big and hard about the pesky problems of running out of energy or irrevocably changing the planet's climate. In fact, this second form of carbon sequestration is so inherently hopeful, in addition to being well presented in the media, that one friend of ours was shocked when we told her that capturing released carbon was indeed a grand idea but nobody was actually doing it yet. She'd let herself believe that it was already an active solution. It most decidedly is not.

But it has to be. It is most decidedly an idea that's time has come. Most of us know this.

Which is why so many of us were outraged when the Bush administrations pulled the plug on clean coal in February of 2008.

FutureGen was to have been the U.S. government's flagship effort in this critical area—a state-of-the-science clean coal plant in Mattoon, Illinois, the prototype operation for all of the clean coal facilities that were to follow, including one in China, for China Huaneng Group, the nation's largest coal-burning utility. But after five years in planning, Energy Secretary Samuel Bodman announced that the Bush administration was yanking funding. The price tag—$1.8 billion—was too high, Bodman indicated. This, we believe, is sadly a classic example of knowing the price of something but having no idea of its cost.

As we transition from the old energy age into the new, we are going to need to rely on coal to provide the energy we need to keep our economies healthy. If it truly is our intent to keep the earth healthy as well—if we're as serious about climate change as we say we are—then we must make the investment in bringing carbon capture technology to its maturity. We must cough up the funds to put the mature technology into the factories and power plants where it can do its work to relieve the earth's carbon dioxide accumulation. And we must—just as soon as the technology is proved—export the technology to China and other emerging economies where coal is the mainstay of their energy mix and share the breakthrough answer to one of the most dire of our energy questions.

A FEW MARKET REALITIES

Power suppliers and distributors provide us with an essential service. Let's start there. For the vast majority of us living in the developed world, life without constant access to an electrical outlet would be, at the very least, a disorienting experience. For some people, like the friend who—truth be told!—actually threw his plug-in electric toothbrush into his backpack to go on a camping trip, trying to function without a place to plug in can feel surreal.

But paying for our electricity is something we're only more or less willing to do. We've become accustomed, however unconsciously, to the subsidies that fix our energy costs below their true market value. A raise in the rates we pay for our utilities can strike us as an infringement on our rights of citizenship rather than as a business decision.

Well, utilities are businesses. They are in the business of making a profit by selling electricity to us, and we have to make sure the policies our governments put into place allow them to continue to function as successful businesses. Otherwise, why should they bother? Lots of us find fulfillment in our work, and a regular paycheck is a big part of the reason why.

As utilities adjust their business models to adapt to the new energy age, we can anticipate some growing pains, perhaps in the form of rate increases. How do we prevent rate increases, and how do we offset the ones we can't prevent, especially as they affect those people who would be most severely impacted by them?

Start with those subsidies. We'll talk again about this subject in the next section of this chapter; here we can settle for pointing out that adjustments in the ways federal subsidy programs are structured would go a very long way in controlling any price increases that might result, even temporarily, in response to the transition to new sources of energy.

The second thing that we can do is to increase the financial stake power utilities have in *saving* rather than in *selling* energy. How do we do that? Provide incentives. Perhaps we provide an incentive for setting up a program in which a utility conducts free energy audits of customers' homes, and then we reward the utility based on kilowatt-hours saved. This is but one scenario that could work, and it's a good one because all of the parties involved—the utility, the customer, the environment, and the economy—are supported in the same incentive plan.

Another idea that could assist utilities in increasing their energy savings is to implement a program that provides their customers with the Smart Meters we talked about in chapter 5 as being so helpful in reducing grid congestion. These meters can help to raise customers' awareness about the time of day and at what rate they are using energy, and they can also help customers to monitor the *amount* of use. Simply becoming more aware of our actions is often the first step in altering them.

Helping low-income families cope with high energy costs should be a goal regardless of utility rates or the sources of energy those rates represent. One way to do this could be through an extension of the home audit program we just mentioned. Providing not only the audit but assistance in acquiring the materials and/or the labor to qualified families to affect the audit recommendations would help to lower a family's energy costs, reach the utility's target for energy savings, and decrease the emissions and consequent environmental impact from energy-inefficient homes. Again, every party to the action wins through the streamlined administration of one initiative.

There is one other idea for helping low-income families that we like very much, especially as it can target homes with children or the elderly, is the institution of what we call "Cold Snap Relief." The way this idea would work is very simple, and easily administered: When the temperature drops below 32° F, pre-qualified households are automatically exempted from energy taxes associated with their monthly heating bill. As people use more energy to heat their home when the temperature drops, Cold Snap Relief kicks in and offers some offset at just the time when a family's energy bill is likely to go up.

FUNDING

At last, we come to the heart of the issue, the make-or-break matter upon which our future energy security rests: Just how are we going to pay for all of these policies and programs? Part of you might be reeling to see the scope of what's required to move us into the new energy age as even as another part of you fully understands that the solution has to match the size of the problem, and we've got a big problem on our hands. Just as the world has never before been as connected and small and *flat* as it is today, it has never before faced a challenge quite the size of this one.

Still, setting policies and making laws accomplishes nothing if money isn't available to fund them. Where is the money going to come from to pay for all that we must do to solve our energy predicament?

Three places.

The first place the money is going to come from is pollution. We need to turn the carbon emissions we create into the cash we need by investing the funds collected through cap-and-trade programs, as well as through green tags and carbon offsets, into renewable energy development and deployment. It's fitting, even poetic, that the source of our problem, CO_2, can become part of the solution.

And though putting cap-and-trade programs into place will be difficult from a political perspective, requiring the tough negotiating skills and thorough resolve that stripping off political blindfolds to see any new vision has always required, accomplishing the second avenue of funding for our renewable future is what is going to take real political will. That second avenue consists of the overhaul of federal subsidy programs— restructuring and streamlining them—and reallocation of available funds.

As a basic matter, we can no longer afford energy subsidies as usual. Thoughtlessly doing what we have always done—which is to feed the lion's share of a limited pool of funds to fossil fuel interests—is no longer simply wasteful and impractical. It's dangerous. Bluntly, the world can no longer afford to subsidize pollution. Subsidy programs—tax incentives, low-interest loans, direct funding of research—have to be redirected to the energy sources that are our future, not the ones that were our past. This restructuring includes such complicated and potentially contentious issues as streamlining subsidies for ethanol, which in the United States currently flow from agricultural subsidy programs, and sourcing them through more appropriate energy subsidy programs.

You can be sure that the current beneficiaries of government subsidies are going to fight hard to retain their interests. But we, and certainly the politicians who are going to have do the fighting on our behalf, have to take the long view. Restructuring subsidy programs now, for all of the complexity and contentiousness involved, comes with benefits beyond even the immediate, critical ones of helping to curb climate change and smoothing the way for the development and deployment of the green technologies that will secure our future energy supply.

In changing our focus on how we create energy from traditional and dirty sources of power to clean and renewable ones, we will reduce pollution. That's just a cut-and-dried conclusion. And reducing pollution means that people will begin to suffer less from pollution-related illness and disease. Less disease means lower health care costs.

Reducing pollution also impacts rising property insurance rates. It's been estimated that weather events in just the next decade will cost $150 billion in insurance payouts. Bringing our energy sources in line with what we know to be climate-friendly practices can help to change the course we are on regarding projected weather-related destruction to our homes and livelihoods. We can avoid the devastation that's surely ahead if we continue along our same, old energy path.

Spending less on things like health care because more people are well, or less on things like insurance or disaster relief because we have taken the steps to mitigate threats to our persons and property, means that even more money is available to invest in other programs. And not just renewables programs either. Education immediately comes to mind as one of the programs that is deserving of a few more dollars than it currently claims. Having more funds available to invest in education, or national security, or middle-class tax relief is part of the benefits in taking a wise and long-term approach now to our energy crisis.

The third place from which the money will come to build green power plants and biofuel distilleries is private investment. As we mentioned previously, it's estimated that private investment will account for up to 86 percent of all the funds necessary to transition to the new energy age.

But private funding hinges upon public policy. It hinges upon federal support for Renewable Energy Standards—the certainty that there will be buyers for green energy products. It hinges upon the streamlining and reallocation of subsidies and creating incentives to green power producers rather than just continuing business-as-usual for fossil fuel sources. It hinges upon carbon pollution coming with a cost—and on the funds that are collected from polluters being directed to green power and fuel projects. Private investors are prepared to step up to the threshold of our energy tomorrow. They are ready—*eager*—to take the better way. It's governments that must now lubricate the hinges and swing open the door to the future.

Taking the long view allows us to see exactly where we're headed if we don't change our energy ways. And it allows us the opportunity to find—and to take—the better way.

Conclusion

THE GREEN ECONOMY

It's SNOWING OUTSIDE AS MICHAEL AND I COME TO THE END OF WRITING THIS book, December. We're looking forward to the holiday season that's approaching as a well-deserved break with family and friends. But two events that we have been following closely during the course of our work are also coming to an end. The first is the United Nations Climate Change Convention on the island of Bali in Indonesia. The urgency of our global climate crisis was restated there in plain language that we can all understand, if we're willing to do so.

The second event is the debate within the United States Congress over the latest iteration of an energy bill. A sweeping bill that encompassed many of the actions we discussed as necessary in the last chapter was whittled to a skeleton of itself in the final draft. A provision that would have required utilities to obtain at least 15 percent of their electricity from renewable sources was dropped, as was a $21.8 billion package designed to extend investment incentives in clean energy by raising oil company tax obligations. The bill might have been the first major step by one of the world's leading industrialized nations in addressing global environmental sustainability and energy security. We find it ironic that in the same week as dire climate change warnings were reissued from Bali, an energy bill was rejected in the United States because many lawmakers saw it as "too ambitious."

Perhaps that's because the critical connection between environmental sustainability and energy security has not yet been fully made in the minds of the people who make the laws. Or it hasn't been made in the minds of *enough* of them to matter to the passage of a realistic energy bill that will lead us to develop a secure, sustainable future. And that's ironic. In poll after poll, election result after election result, people—you and I—are telling our lawmakers that these are two of our primary concerns. We get it. Why don't they?

The continued stalling on these two grave matters may be the result of a failure to connect the two *Es*—environment and energy— but it may also be the result of a failure to incorporate a third, crucial *E*—economy—into the equation.

We think it's time, therefore, to definitively reframe the issues of climate change and our growing demand for energy as what they really are: threats to our global economy. In not addressing the first two *E*s now with bold, sweeping action, we continue to place our vulnerable coastlines at risk to more disastrous hurricanes. We continue to place our stressed grids at risk to more, and more frequent, brownouts. We continue to play a dangerous game of chicken by daring dirty, dwindling fossil fuel sources to keep pace with our increasing consumption of them. In the end—from the costs of rebuilding lost cities, to the losses in productivity caused by brownouts, to the soaring costs of health care for pollution-related disease—these are all *economic* gambles.

Renewable Energy + Increased Energy Conservation + Energy Technologies That Don't Exist Yet = A Secure Energy Future

It's a simple formula. Making it work is, clearly, more complex. That's why, in reframing the issues of environmental sustainability and energy security in their broader context, as real threats to our national economies, we think it's also appropriate to reframe the call to action on these issues in terms of patriotism. It's unpatriotic to leave the citizens of your country vulnerable to natural disaster. It's unpatriotic to leave the citizens of your country dependent on sources of energy that sap both their health and their earning power. It is most unpatriotic to put your country in the position of being ill prepared to meet its economic future.

The global economy will, sometime in the not-so-distant future, begin to turn on renewable energy. It will because it will have to. Fossil fuels will remain in the energy mix for some time, of course, because our growing demand for electricity will require that. But renewable sources of power and fuel will become more and more commonplace and conventional, filling ever more of our energy needs. Some countries will adopt sustainable power sources sooner than others. Those countries will emerge as world leaders in the coming green economy. Where do you want your country to stand? If it's in a leadership position, you must demand that your current leaders act.

Those countries that don't act will be at a serious economic disadvantage.

That's as simple as our complex environmental-energy-economic challenge gets.

As is nearly always the case, what is blocking the way of such patriotic progress is a way of thinking about our contemporary challenges that is rooted in the past. In order to streamline the complex process of moving forward toward secure and sustainable environmental, energy, and economic policies we need leaders who are both willing and able to throw off the lethargy of "what we have always done" and embrace *what is possible*.

And what is that? What is possible?

Economies that continue to run on fossil fuel energy are not possible because supplies of those environmentally toxic fuels are dwindling. A "hydrogen economy" is not possible because the technology is at least fifteen years from maturity—and even if it becomes mature, the construction of a wholly new infrastructure to support it will be too costly to be practical. A resurgence in the use of nuclear energy on a grand scale is not possible because the associated health risks are too grave—and because the cost of minimizing those risks is simply too high to sustain a cost-benefit ratio that industry can embrace.

But is it possible to meet, at minimum, a quarter of our global transportation needs with biofuels by the year 2020? With policies in place that support the development and production of cellulosic ethanol, and that remove the prohibitive tariffs now associated with importing sugar ethanol, well, it will be a stretch, challenging us to put our money where our resolve is, but yes, it is possible.

Is it possible that the people of Africa have a new and vital role to play in our energy future through the cultivation of a hardy weed that already calls the continent home? Is it possible to uplift people out of crushing poverty at the same time that we relieve the poverty of our fuel supply? With the right investment incentives—and a vision that balances agricultural practices with environmental wisdom—yes, it is.

Is it possible to bring to fruition the long promise of clean coal technologies? To deploy this technology in a cost-effective way so that people around the globe are no longer stricken with life-threatening pollution-related illnesses—and so that the United States and China, instead of competing as the world's largest carbon emitters, can sustain and grow their economies in healthy competition? Is it possible to build enough solar fields and wind farms and geothermal plants and hydroelectric dams that we can begin to retire the world's inefficient coal plants altogether? With strong government support for private investment, yes, it is.

Is it possible that instead of brushing my teeth with a toothbrush made from chemical materials, I'll start the day with a toothbrush made from plant matter? That instead of packing my daughter's lunch in a container made from petroleum-based plastic, I'll send her off to school with a lunchbox made from sugar? That rather than tossing the morning newspaper into the recycling bin after we're finished reading it, Michael and I, and all of you too, will be sending our trash off to a plasma converter so that Saturday's sports section can be turned into the electricity we'll need to watch Sunday's game on TV? By investing wisely in the development of by-products from biofuel operations—and by supporting those technologies that don't quite exist yet—yes, it is.

But what is possible depends on how you see the future. If you believe a problem is insurmountable, it probably is because you won't possess either the will or the faith to solve it.

If, however, you posses a vision—a vision as strong as FDR's when he mobilized the private sector with tough and effective government policies and turned on America's lights—then our secure and sustainable energy future is possible.

SOURCE NOTES

OUR READING FOR THIS BOOK CONSISTED OF LITERALLY THOUSANDS OF BOOKS, reports, white papers and articles from periodicals as diverse as the *New York Times* and the *Los Angeles Times* to *Xinhua Business Weekly* and the *Shanghai Daily*, Shanghai's daily English-language newspaper. Only sources that have been directly quoted or are significantly relevant to the flow of ideas contained within this book are cited. Where sources are cited within the text, citations are not repeated here. Additionally, our resources included interviews with some key players on the renewables stage; these interviews are cited within appropriate chapter notes except where sources declined to be named.

1 GREEN MONEY

The U Street merchants came to our attention by way of Shawn G. Kennedy's article in the Washington Post ("Washington's Small Businesses Tap into Green Power," August 15, 2007). Further information was furnished by Nazim Ali, Ayari de la Rosa, Andy Shallal, and Gary Skulnik. The history of U Street was derived from information available through Cultural Tourism DC in collaboration with the Historical Society of Washington, D.C. and the Downtown DC Business Improvement District. We've drawn from the work of Ernest J. Moniz and Melanie A. Kenderline (notably "Meeting Energy Challenges: Technology and Policy," Physics Today, April 2002) in this and subsequent chapters for our discussion of the present global energy challenge. The article "PetroProzac, Dasani-style" by Dave Cohen (ASPO-USA/Energy Bulletin, September 5, 2007) inspired our choice to focus on plastics as an example of energy use. Additional sources of materials and statistics were the International Bottled Water Association (IBWA), the Beverage Marketing Corporation (BMC), the National Park Service, the American Chemistry Council, and the EPA.

2   ENERGY PRESENT

The background for the history of fossil fuel use, as well as technology development and contemporary statistical energy consumption information, was drawn from materials available through the U.S. Department of Energy (D.O.E.), the U.S. Department of Fossil Energy, the Energy Information Administration, the International Energy Agency (IEA), the American Council for an Energy-Efficient Economy, the California Energy Commission, and CAISO (California Independent System Operator), which operates the majority of California's high voltage wholesale power grid. Two articles from National Geographic ("Carbon's New Math," Bill McKibbon, October 2007, and "The Case of the Missing Carbon," Tim Appenzeller, February 2004) were particularly helpful with regard to carbon and terrestrial carbon sequestration data, as was the work of Janet Cushman, Gregg Marland, and Bernhard Schlamadinger of the Oak Ridge National Laboratory. Information about mercury, sulfur, and other pollutants was drawn from EPA sources. Additional materials about fossil energy consumption and supply status was drawn from the testimony of Daniel Yergin to the Committee on Foreign Affairs, U.S. House of Representatives hearing on "Foreign Policy and National Security Implications of Oil Dependence," March 22, 2007. Material on nuclear energy was drawn from D.O.E. testimony before the subcommittee on Energy and Air Quality, Committee on Energy and Commerce, U.S. House of Representatives hearing on nuclear power, March 27, 2001; the sources of statistical data were the International Atomic Energy Agency (IAEA) and the U. S. Nuclear Regulatory Commission (NRC). The quotes by President Dwight D. Eisenhower were taken from his address to the 470th Plenary Meeting of the United Nations General Assembly, December 8, 1953. The source for background on the R.E.A. was "Rural Electrification Administration," Laurence J. Malone, a professor of economics at Hartwick College and a Carnegie Foundation Fellow, published by the Economic History Service, a service of the Economic History Association.

3 RENEWABLE ENERGY, A PRIMER

Statistical information about the Christmas tree farming industry was drawn from data available through the National Christmas Tree Association and the British Christmas Tree Growers Association. Material about holiday electrical use was taken from the D.O.E.; the article "Efficient Lights Can Brighten Holiday Season," by Ken Scheinkopf, Augusta Chronicle, October 22, 2006, provided comparative statistical information. Statistics regarding power usage were drawn from the International Energy Agency. Information on hydrogen technology was drawn from various sources including: "Is There Hope for Hydrogen?," Joan Ogden, Daniel Sperling and Anthony Eggert, American Chemical Society Energy Bulletin, October 10, 2004; "Hydrogen and Fuel Cells: Pathways and Strategies," Joan Ogden, presented to Briefing on Future Transporation Energy Opportunities and Challenges, Washington, D.C., January 23, 2007; "Potential Environmental Impact of a Hydrogen Economy on the Stratosphere," Tracey K. Tromp, Run-Lie Shia, Mark Allen, John M. Eiler, Y. L. Yung, Science Magazine, June 13, 2003; and "Hydrogen Economy Fact Sheet," a White House news release, 2003.

4 POWER PLAYING

Sources for community renewable initiatives include: "Community-scale projects in Europe beat sustainable energy and climate change targets," Kimberly Conniff Taylor, International Herald Tribune, November 13, 2007 and the William J. Clinton Foundation. Costs and current consumption figures for renewable technologies was drawn from the IEA, the American Council on Renewable Energy (ACORE), the D.O.E., the D.O.E.'s office of Energy Efficiency and Renewable Energy, and the California Energy Commission. Background on renewable technologies was provided through the D.O.E. and augmented by various sources. For technical background information we are particularly indebted to the editors and contributors of Renewable Energy World Magazine, www.renewable-energy-world.com.

Solar: Berkeley, California solar roof initiative, City of Berkeley web site, www.ci.berkeley.ca.us, and "Good News: Solar Energy Breakthrough Brings Green Power Closer," John Vidal, The Guardian, December 29, 2007; military use of renewable energy sources, "In the Iraqi war zone, U.S. Army calls for 'green' power," Mark Clayton, The Christian Science Monitor, September 7, 2006; information on asphalt road surfaces as solar fields is from "Design Tool for the Thermal Energy Potential of Asphalt Pavements," Marcel Loomans, Henk Oversloot, Arian de Bondt, Rob Jansen and Hand van Rij, as presented at the Eighth International IBPSA Conference, Eindhoven, Netherlands, 2003.

Wind: Background information on windmills and the history of wind power was drawn from articles posted on www.windmillworld.com and from Darrell M. Dodge's "Illustrated History of Wind Power Development" at www.telosnet.com; additional sources included The Wind Power Monthly, the wind industry's monthly publication, and material available through the Sierra Club, the Utility Wind Integration Group (UWIG), and the Global Wind Energy Council; wind energy storage material is from "Compressed air wind energy storage," Bryce Finley, Energy Bulletin, November 27, 2005. Geothermal: Additional materials were drawn from information available through Iceland American Energy, Inc.

Hydroelectric: Additional materials were drawn from the Georgia Water Science Center and from the Green Trust for Sustainability and Renewable Energy.

5   POWER TO THE PEOPLE

Overviews of the workings of electrical grids were provided through Energetics, Inc., sponsor of the 2008 National Electricity Delivery Forum; the American Institute of Physics; American Science Magazine, a publication of Sigma Xi, the Scientific Research Society, from the article

"The Electric Transmission Paradox," Vito Stagliano and Jolly Hayden, The Electricity Journal, March, 2004, and from the report "Inducing More Efficient Ethanol Production," Thomas R. Casten, May, 2007. For statistical information we relied, again, on the IEA and U.S. government agency data. The quote attributed to Spencer Abraham was taken from a report by the American Institute of Physics.

6 FUELISH CHOICES

Background material on the history of biofuels was drawn from several sources including: The Ohio Energy Project, Yokayo Biofuels, and Energy Intelligence, an online energy reporting service. Greg Bafalis, CEO of Green Earth Fuels, provided background and insight concerning biofuel transportation and integration into traditional fuel supplies. Background information on biomass and biomass statistics was accessed through the National Renewable Energy Laboratory (NREL), and studies completed at the Oak Ridge National Laboratory. Background and statistical information about ethanol was drawn from a variety of sources as well, including Renewable Fuels Association (RFA) documents and the audio interview, "The Ethanol Report," with Bob Dinneen, chair of the Cellulosic Ethanol Summit; "Fuel Without the Fossil," Matthew L. Wald, New York Times, November 11, 2007; "Fuel Ethanol: Background and Public Policy Issues," Congressional Research Service Report for Congress, December 2004; "Growing Energy," a report by Nathanael Greene, for the National Resource Defense Council, December 2004. Information about Brazil's ethanol transition that was particularly helpful was drawn from "Sugar Rush," Antonio Regalado and Grace Fan, the Wall Street Journal, September 10, 2007; and "The Brazilian Sugarcane Ethanol Experience," Marcus Renato Xavier, in a paper for the Competitive Enterprise Institute, February 15, 2007. Carbon dioxide reduction statistics were drawn from "Well-to-Wheel Energy Use and Greenhouse Gas Emissions of Advanced Fuel/Vehicle Systems," the work of Norman D. Brinkman,

General Motors Corporation, and from "The Debate on Energy and Greenhouse Gas Emissions Impacts of Fuel Ethanol," Michael Wang, Argonne National Laboratory, Center for Transportation Research, Energy Systems Division Seminar, August 3, 2005. Information about land management analysis for biofuel crops was drawn from "Producing Cellulosic Feedstocks from Currently Managed Lands," Lee R. Lynd, October 7, 2005. Information about biofuel production costs was culled from various articles that have appeared in The Economist magazine, and from information available through the D.O.E., the UK Petroleum Industry Association, and New Energy Finance, a company that specializes in providing renewable investment analysis. Additional information on cellulosic ethanol was drawn from "Oregon Cellulose-ethanol Study," prepared for the Oregon Office of Energy by Angela Graf and Tom Koehler, June 2000, and from "Use of U.S. Croplands for Biofuels Increases Greenhouse Gases Through Emissions from Land Use Change," Timothy Searchinger, Ralph Heimlich, R. A. Houghton, et al, Science Magazine, January 28, 2008. Some information about the transition of the transportation sector was drawn from the White Paper "Transforming Global Transportation, Fuel independence at country level as a business opportunity," Shai Agassi and Andrey Zarur, December 2007. Quotes from Senators Max Baucus and John Tester were drawn from a Sustainable Oils news release, November 20, 2007.

7   CREATING A RENEWABLE INDUSTRY

Background information about worldwide energy subsidies was drawn from information available from the Global Subsidies Initiative. Information about fossil fuel subsidies was drawn from "Money Down the Pipeline: The Hidden Subsidies to the Oil Industry," a report issued by the Union of Concerned Scientists in 1995; "Summary of the economic and social survey of Asia and the Pacific," report to the United Nations Economic and Social Council, April 2006; "The Oil Kingdom at 100: Petroleum Policymaking in Saudi Arabia," a report by the Washington Institute for Near East Policy, 2007, as well as available national budgetary data from relevant government sources. Information about the carbon market was drawn from documents accessed through the EPA; the D.O.E.'s office of Energy Efficiency and Renewable Energy; the Clinton Foundation, the Clinton Climate Initiative; and treehugger.com. Further information was drawn from a report by the Ernest Orlando Lawrence Berkeley National Laboratory, "The Treatment of Renewable Energy Certificates, Emissions Allowances, and Green Power Programs in State Renewable Portfolio Standards," Edward A. Holt and Ryan H. Wider, April 2007, and from "State and Trend of the Carbon Market 2006," a report issued jointly by the International Emissions Trading Association and the World Bank, Karan Capoor and Philippe Ambrosi, authors. Information about the success of the sulfur cap-and-trade program was supplied through the Environmental Defense Council. Articles in various periodicals also supplied important background information and insight including "Cleaning Up Carbon," John Goff, CFO Magazine, April 2007; and the opinion pieces, "A Carbon Cap That Starts in Washington," Judith Chevalier, New York Times, December 16, 2007, and "We need to bring climate idealism down to earth," Lawrence Summers, Financial Times, April 30, 2007. The quote attributed to Mayor Bloomberg is taken from eNewsUSA, a publication of Waste Information & Management Services, Inc, November 6, 2007.

8   TECHNOLOGIES THAT DON'T EXIST YET

The authors are grateful to Lloyd Spragins and Jeff Voorhis for interviews. Further information on Plasma Converters was accessed through www.startech.net. Information about jatropha was culled primarily from www.jatrophaworld.org, and news coverage of the BP and D1 Oils joint partnership to develop jatropha-based biofuel, June 2007.

9   THE GLOBAL LOW-CARBON ECONOMY

Information about Gordon Brown was drawn from BBC reports. Information concerning activities at the U.N. Climate Conference in

Bali, as well as Kyoto, its predecessor conference, was drawn primarily from U.N. postings as the conference was in progress as were quotes from Mr. de Boer. Details about China's renewable energy plans and funding was drawn from various newspaper accounts, notably "High Oil Prices make Asia pursue green energy," David Fogarty for Reuters, September 9, 2005, and information available from China's Energy Research Institute. Much of the Abu Dhabi material was drawn from articles issued by Spie, an international science society, including the Peter Evan quote, as well as *Commodity Risk*, a quarterly publication dedicated to risk management and commodity trading. The Technology Development Program at M.I.T. provided additional background material on the Masdar Initiative.

10   THE GREEN STANDARD—BEST PRACTICES

Details of Saudi Arabia's renewable initiative are drawn from the report, "Renewable Energy Potentials in Saudi Arabia," S. A. M. Said, I. M. El-Amin, A. M. Al-Shehri, King Fahd University of Petroleum & Minerals. The paradox of conservation vs. increased consumption is addressed in the article "Energy savings lost to 'efficiency paradox,'" Shawn McCarthy for the *Globe and Mail*, November 28, 2007. Interviews with Rob Watson contributed much value to our discussion of energy conservation.

INDEX

222–4, 234. *See* "clean" coal; strip mining
"Cold Snap Relief," 226
communications technologies, 17–18, 24, 43
compressed air energy storage (CAES) systems, 110–11
condensed solar power (CSP): *See* solar thermal electricity
conservation: *See* energy conservation
corn, 111, 136, 138–42, 144, 146–7, 151–6, 215
Corporate Average Fuel Economy (CAFE) standards, 212
crop rotation, 131, 155

D

Denmark, 37, 60, 77–8, 95
Diesel, Rudolph, 132–4, 186
diesel fuel oil, 44, 113–14, 132–5, 139, 141
direct heating systems, 104–5
Dispatchable Wind Power System (DWPS), 110

E

economic incentives, 6, 16–19, 22, 42, 85–8, 96, 145–6, 151, 162, 214, 216, 221–2, 225,

227–8, 231, 234.
See subsidies
Eisenhower, Dwight D., 52, 103
El Hierro, 77
electric cars, 148, 165
electricity, 23–4, 31–7, 39, 42–3, 53–5, 64–5, 71, 75–7, 79, 82–6, 89–101, 103, 106–14, 117–24, 139, 144, 154, 156, 167, 172–5, 177–8, 188–9, 197, 214, 217–19, 224, 231–2; distribution, 117–24; global demand for, 23; price of, 120; shelf life of, 122; storage, 101, 107–11; transportation of, 114. *See* clean electricity; electric cars; electrification; grid system; hertz; power plants; solar thermal electricity
electrification, 32–4, 56, 83, 94–5, 99–100, 211
Electronic Arts Inc. (EA), 75–7. *See SimCity*
energy, 4–6, 8, 23–4, 29–36, 42, 46–9, 55, 61–6, 68–71, 75–9, 81–5, 90, 96–9, 103–4, 111, 120, 126, 142, 156–7, 161, 184–5, 205–9, 225–6; definition of, 31, 62; evolution of, 29–34; units of, 63–4. *See* clean energy; energy conservation; energy